RURAL AQUACULTURE

RURAL AQUACULTURE

Bruno Augusto Amato Borges

www.delvepublishing.com

Rural Aquaculture

Bruno Augusto Amato Borges

Delve Publishing

2010 Winston Park Drive,

2nd Floor

Oakville, ON L6H 5R7

Canada

www.delvepublishing.com

Tel: 001-289-291-7705
       001-905-616-2116

Fax: 001-289-291-7601

Email: orders@arclereducation.com

ABOUT THE AUTHOR

Bruno Augusto is Aquaculture Engineer from Federal University of Santa Catarina. His focus is on writing projects and development of new technologies in aquaculture. He is currently project service provider, consultant and develops research in the field of Biofloc Technology (BFT) for freshwater fish farming.

TABLE OF CONTENTS

LIST OF ABBREVIATIONS

CARD	Coalition for African Rice Development
DC	Direct Current
EMS	Early Mortality Syndrome
FAO	Food and Agriculture Organization
FSH	Follicle Stimulating Hormone
GDP	Gross Domestic Product
GnRH	Gonadotropin Releasing Hormone
IMTA	Integrated Multi-Trophic Aquaculture
ISA	Infectious Salmon Anemia
MDG	Millennium Development Goal
MSA	Magnuson-Stevens Fishery Conservation and Management Act
NOAA	National Oceanic and Atmospheric Administration
NOFA	Near- and Offshore Finfish Aquaculture
NPDES	National Pollutant Discharge Elimination System
SHEP	Smallholder Horticulture Empowerment and Promotion Project
SHEP PLUS	Smallholder Horticulture Empowerment and Promotion for Local and Upscaling
SRFI	Sustainable Rice Fish Integration
TICAD V	Fifth Tokyo International Conference on African Development
U.S.	United States
USAID	The United States Agency for International Development

PREFACE

The field of aquaculture is a rapidly growing sector on the food industry. Aquaculture has been instrumental in meeting the demands of the world population in terms of food and nutrients and has been contributing heavily towards the populational growth.

Aquaculture has been seen as a profitable industry in the urban as well as the rural areas. The urban regions use aquaculture to provide themselves with the animal-based food, so that they can have access to better nutrients. This has been on the rise, catering to the increasing population and increasing earnings of the people. Aquaculture has also supported the population by making room for the new employment opportunities and creating jobs in places where other kinds of industries are either not found or difficult to be established. The aquaculture sector has supported the livelihoods of the people living in or near the coastal areas in an emphatic manner and has become a significant part of their lives.

The rural regions use aquaculture as a source of food to provide themselves with the basic forms of food available to them, which also can help them grow and develop and can additionally provide them a primary or an alternative form of income. This book brings the focus of the readers to the rural aquaculture and explains the various aspects related to it in detail.

The book introduces rural aquaculture to the readers and explains them the fundamental elements related to it. It explains them the production framework in rural aquaculture and informs them about the intensity of production in this sector. The readers are informed about the changing points of view in regard to the rural aquaculture. Also presents the possibilities of integrating rural aquaculture with agriculture. It discusses the various synergies in the integrated aquaculture-agriculture systems and informs the readers about the various benefits that they may have by adopting the process. It also talks about the various considerations that must be taken in the process. Then, informs the readers about the various production systems in the rural aquaculture and discusses with them, the sustainability in food production. The readers are informed about the various production facilities in the world and the species that are cultured in the modern production systems. The following chapter informs the readers about the various technologies that may be employed in rural aquaculture and discusses the various kinds of approaches to the aquaculture technologies.

Moving further, discuss the impact of rural aquaculture on the livelihood of the people involved in it and the food security of the world. The readers are informed about the various multipliers from the economic and employment point of view and are told about the revenues generated from trade and taxation.

It lists various techniques that are employed in biotechnology to be used in rural aquaculture. The chapter also discusses the challenges related to the application of biotechnology. Then, explain the various economic aspects of rural aquaculture and also the social aspects of it. The chapter explains the effects of various socio-economic factors on rural aquaculture and discusses the drivers that drive the economy and social aspects of the sector. Also, discusses the various issues and opportunities in rural aquaculture by first bringing the focus of the readers to the important issues in the sector. The readers are informed about the challenges that arise in the field of rural aquaculture and the various opportunities in the aquaculture sector.

The book ends by discussing the various prospects of aquaculture in the future, talking about the sustainability in the field of aquaculture that can be achieved in the future and the various threats and weaknesses that are there in aquaculture. It also talks about the manner in which the growth of aquaculture can be achieved to the desired degree and the ways rural aquaculture can be developed in the future.

1

Introduction to Rural Aquaculture

CONTENTS

Aquaculture is concerned with farming of animals and plants in inland and marine water. This chapter concentrates of rural aquaculture. It describes what rural aquaculture encompasses. The chapter describes production framework of rural aquaculture. It then discusses intensive practices of aquaculture which can improve yields in rural areas. The chapter finally highlights the changing perspectives of rural aquaculture covering aspects like innovations by rural farmers, the establishment of ornamental fishing villages, and gender mainstreaming.

1.1. INTRODUCTION

Aquaculture consists of diverse systems of farming animals and plants in the inland and marine waters. It is a significant economic activity. Aquaculture becomes an attractive and important component of rural livelihoods in situations where increasing population pressures, environmental degradation or loss of access limit catches from wild fisheries.

Rural aquaculture refers to farming of aquatic organism undertaken by small-scale farming families or groups in the society. They generally have a limited resource base and employ usually by extensive or semi-extensive techniques of farming. Many of the rural aquaculture activities are carried out for self-sustenance. The level of fish production is generally low, and the only source of family income.

The resource-poor base of most farms requires off-farm agro industries inputs to intensify production. This means that they mostly rely on inorganic feed instead of formulated feed. This enables them to market the feed at a lower cost and make it affordable to poor customers. The growth of aquaculture can be achieved in two ways. First, by increasing the area on which aquaculture is done. Second, by intensifying the production in the same area. When compared to agriculture, there is considerable potential for horizontal expansion of agricultural land. Aquaculture makes use of saline soils mangroves, swamps which cannot be utilized for agriculture. Aquaculture activities can also be carried out on rivers, reservoirs, natural, and man-made lakes.

The recent studies on rural aquaculture have found that aquaculture cannot be viewed as an isolated economic activity. It is an important contributor to rural development and is an aspect of holistic approach to advancement of the rural areas. Researchers have proved ho rural aquaculture can transform the lives of the underprivileged. It offers various benefits like offering a cheap source of nutritious food, generates jobs, reduces disparity in income.

Figure 1.1: Representation of small-scale aquaculture in rural areas.

Source: https://upload.wikimedia.org/wikipedia/commons/b/b0/Small_scale_ aquaculture_in_Kenya.jpg

1.2. PRODUCTION FRAMEWORK OF RURAL AQUA-CULTURE

The production framework consists of three interrelated aspects: cultured species, culture facilities and husbandry.

The choice of facilities is dependent on the cultured species and similarly the facility determines the species that can be cultured. The husbandry in turn is dependent on culture facilities and species.

1.2.1. Cultured Species

Over 200 species of fishes are farmed both in marine and inland water areas. The most common species in rural aquaculture are mollusks and finfish and to some extent, shrimps and prawns along with seaweeds and aquatic macrophytes.

For rural aquaculture, herbivorous and omnivorous (mainly tilapia and carps) are significant and constitute a major portion of inland finfish production. Other specie which dominates the coastal aquaculture are herbivorous milkfish. Culture of tilapia can also form an important part of the coastal rural aquaculture.

Some nations practice culture of small-scale seaweed. Freshwater macrophytes which is significant to rural aquaculture receive little consideration from scientists. In Asia, water spinach is an important in rural aquaculture which experiences a humid climate.

Majority of the farmed aquatic organisms are tamed but not domesticated genetically. In other words, they are similar to the wild types. This presents a huge potential which has not been exploited yet.

1.2.2. Culture Facilities

Many cultural facilities are used in rural aquaculture. Some of the cultural facilities which houses aquatic organisms seasonally include, ponds, rice fields, irrigation ditches. These may be open to allow significant water exchange with the surrounding environment at least during the rainy season. Rural aquaculture is mostly practiced in ponds.

They are commonly found in floodplains where farmers dig them minimize risk of flooding. Small ponds are useful for trapping wild fish. In many countries, construction of ponds in rural areas is becoming very popular due to unreliable water supply.

Figure 1.2: Representation of pond construction for aquaculture.

Source: https://upload.wikimedia.org/wikipedia/commons/e/ef/Pond_Construction.jpg

Pens and cages are used for keeping the fish. These enclosures allow exchange of water with the surrounding water body. These kinds of enclosures are mostly used by wealthier farmers. This is because water exchange requires nutritionally formulated feeds which are costlier. The small-scale farmers rely on hapa, a kind of enclosure made of nylon which is used for nursing fish seed.

Sedentary aquatic organisms like mollusks, seaweeds and macrophytes are cultured in open facilities. Hard substrates like wood, rock or concrete are used in this case. They are either placed at the bottom or are used to suspend the cultured a species in their enclosure. Clams prefer soft surfaces like mud and sand.

1.2.3. Husbandry

Aquaculture husbandry consists of three stages: seed production, nursery and grow-out. Capture fisheries help in obtaining seeds of some species from natural water bodies. These species, unlike farmed ones, do not have closed life cycle. Hatcheries are the facilities where modern techniques are used to produce seeds. Hatcheries are suitable for early nursing, however, for advanced nursing, separate enterprises are more cost effective.

Two of the largest aquaculture systems in the world are Chinese carp polyculture and Indian major carp. They are mostly dependent in wild seed from major rivers. Since the 1960s they have become almost totally dependent on hatchery-produced seed as induced spawning techniques involving hormone injection have been developed. In East and Southeast Asia, environmentally induced spawning of common carp has been used since long. However, one major challenge in rural areas of Asia, Africa and Latin America is unavailability of seed. In South East Asia, the rural farmers mostly depend on the local carnivorous species, snakehead, and walking catfish. However, this practice has not been very successful because of their feeding habits.

The three stages of husbandry are dependent on factors like stocking and harvesting strategies, water quality management, feed quality, vulnerability to diseases and medical facilities. Both monoculture and polyculture are popular practices. Monoculture is useful when a single, high value species has to be intensively cultured. Such species are often fed formulated feed.

In rural aquaculture, polyculture is more popular, since more than one species is able to benefit from different feeding niches of extensive and semi-intensive systems in which natural food predominates. Both harvesting and stocking can be practiced either in single or multiple. However, to achieve highest efficiency, multiple harvesting and stocking is preferred. It utilizes the feed and space better. In rural areas, extensive and semi-extensive scales of production prevail. Hence, natural food and supplementary feed are more popular. They are produced in situ, and supplement nutritional inputs are not used. In semi-intensive, fertilizers are added. Moreover, uneaten feed and fish faces expedite the growth of natural feed.

Natural food like benthos and plankton are rich in protein. The supplementary food enriches the natural food further. They are mostly made of carbohydrate rich food like tubers and brans, which are less expensive. The fertilizers used could be organic or inorganic. Agricultural by-products like waste vegetables, broken rice, food waste is used as supplements.

Green fodder is also used, which may be wild or cultivated terrestrial vegetation. Examples of such vegetation are macrophytes, grass etc. Similarly, agro-industrial by-products are also used, such as oil cakes. Additional formulated feed may be required when fertilization alone is not enough to sustain the culture facility.

Water quality management is focused on water inside the culture facilities, though water exchange with surrounding environment is common. In case of open water culture systems, have a more intimate relationship with their surroundings, since there is hardly any demarcation between the two. As culture systems become more intensified, the proportion of dissolved oxygen in water reduces and metabolic by-products like ammonia increases. This is detrimental for the growth and survival of the organisms.

Rural aquaculture systems are rarely constrained by over-fertilization and overfeeding but on better endowed farms high productivity levels can be sustained in static water ponds through balancing nutritional inputs and water quality. In most of cases, does not require mechanical aeration, which is common in intensive and superintensive practices. Disease is more prevalent in intensive systems. The high density of cultured organisms is one of the stress factors and increases the susceptibility of hatcheries and grow out systems to diseases.

1.3. AQUACULTURE PRODUCTION INTENSITY

The majority of the aquaculture products are derived from extensive to semi-intensive aquaculture systems. When extensive farming is practiced, it generally involves unsophisticated methods, uses natural food. The yield from such practices is generally low. When the production intensity is increased, fish are stocked using deliberate efforts. Organic, inorganic fertilizers and low cost supplemental feeds are used to improve natural food supply.

The system found most frequently is the farming of fish in ponds, however rice-fish farming or the stocking of fish into natural or impounded water bodies are also included as aquaculture systems. The production from this type of aquaculture production cannot be estimated accurately. This is because small-scale and dispersed production data is not found in official records. In most cases the produce is traded or consumed locally.

1.3.1. Risks and Benefits

The activities that have a positive impact on the rural population include fry nursing, improving nursing network, integration of aquaculture with agriculture, sustaining and restoring aquatic biodiversity using simple management methods.

In the coastal regions, many rural families gain employment by farming oysters, fish, shrimps, mussels, seaweeds etc. Intensive aquaculture systems having higher output to input ratio. They make use of technology and employ better management control. Generally, facilities are set up which are operated with higher stocking densities. They make use of manufactured feed and chemotherapeutant intervention in order to achieve this.

The remote rural regions in the Americas and Europe practice intensive inland and coastal cage aquaculture specially for salmonids, which have a high economic value. In Australia and Asia as well, similar practices are encouraged for warm water fishes like snappers, yellowtail and sea bass. Coastal shrimp farming has high value, especially for export, which generates foreign exchange. Hence, it is becoming increasingly popular.

Though cash economies of many coastal regions have improved because of development of aquaculture in those regions, there has been various environmental and social impact caused by some forms of aquaculture. There is an increased focus on these aspects in order to remediate the situation.

The importance of aquaculture in the rural areas is mainly because it is a source of high nutrition food, generates employment and contributes to improving farm sustainability. Aquaculture in small farmer systems provides high quality animal protein and essential nutrients, especially for nutritionally vulnerable groups, such as pregnant and lactating women, infants and pre-school children. The weaker section of the society can access proteins at affordable prices. It creates an opportunity for self-employment, jobs for women and is a source of revenue by selling high value products. Larger farms provide employment opportunities, in seed supply networks and support services.

The indirect benefits of intensive production are that plenty of fish is available locally in rural and urban markets. It also provides opportunity for increased income from revenue of other farm products that generate income, which becomes available because of higher consumption of fish. Aquaculture can also benefit the landless from utilization of common resources, such as finfish cage culture, culture of mollusks and seaweeds, and fisheries enhancement in communal water bodies).

Another significant contribution of aquaculture, which is particularly relevant for integrated agriculture-aquaculture systems is their contribution making farm production more efficient and ensure sustainability. The by-products like manure from livestock and crop residues can be used as feed inputs for small-sized aquaculture farms.

Fish farming in rice fields helps in integrated pest management as well as managing vectors of human medical importance. The fishing ponds are also used as water reservoirs for irrigating the agricultural lands in areas which experience water shortages.

There are innumerable benefits if aquaculture because of which it has grown rapidly since 1970s. It has been the fast-growing sector in food production in several countries in the recent times. It has achieved growth of 11.0 % per annum since 1984. This is much bigger compared to the 3.1% for terrestrial animal meat. In the year 1999, the production of all cultured aquatic organism was estimated at 42.8 million mt.

All aquatic organisms cannot be cultured, however, the species identified as culturable is increasing. According to FAO's recent researched, the demand for and supply of fish and its products will increase significantly in the coming years. There are many factors which will contribute to this increase like population growth, economic development and changes in consumption pattern.

In most countries, production from capture fisheries is expected to remain the same or decrease since the maximum yield has been attained. The inland fisheries may yield more fishes; however, sustained efforts are required which is not an easy task.

Inland, fisheries are at more risk of experiencing environmental degradation like destruction of watershed, development of water control structures. These are more relevant for the changing rural environment. Thus, aquaculture has to play an important role in the rural economy and its growth is expected to continue in the future.

Figure 1.3: Representation of extensive aquaculture farming.

Source: https://upload.wikimedia.org/wikipedia/en/7/7c/Fish-farming-viet-nam.jpg

1.3.2. Aquaculture Intensification and Expansion

In order to continue the trend if increased production through aquaculture, either more areas have to be obtained for practicing farming, or improved technology has to be used for intensifying the production. There are several generic technologies which can contribute to this intensification process. The major constraint in achieving these are the socio-economic and institutional issues.

It will be beneficial if land-based culture system is developed in inland areas. Aquaculture can be practiced in the same land as agriculture. The aquaculture tanks can be used for irrigation purposes. Similarly, many strips of land which are not suitable for agriculture, like saline marsh areas and swamps can be used for aquaculture.

There are several kinds of water resources which can be used for aquaculture, like the rivers, lakes, estuaries, mangroves, coral reefs that are suitable for integration of well-controlled, sustainable aquaculture, enhancement or other form of aquatic animal management, into rural development. In order to increase yields feeds or fertilizers have to be used. These can be derived from the agricultural land or even off-farm sources.

Development of infrastructure can help in minimizing external cost which is incurred in feed and fertilizer. This allows the farmer to intensify the production. This is dependent on many factors like investment in production system, availability of finance and access to developed markets.

As mentioned previously, many of the technical aspects of aquaculture are relatively well developed, however there is a knowledge gap between what is known globally and what is available to farmers. The rural extension systems are weak and there are generally few examples of intensified local aquaculture which the farmers can emulate. As a result, they are not very open to take risks are innovate new processes.

Biotechnology offers various tools and techniques which can help the farmers to improve the quality of feeds, increase the production of farmed species, conserve the environment and restore it, and make the management process very efficient. Genetic improvement programs can help in improving yield. Selective breeding programs have been successful in yielding consistent gains between 5–20% per species.

Biotechnology helps in improving the breeding capabilities, improve the larval nutrition and allow genetic manipulations to be performed on many aquatic species of plants and animals.

The breeding programs are particularly significant for restocking water bodies which have endangered species of aquatic organisms. At present, the biotechnological tools are very expensive. They are mostly developed for farming systems with high inputs of feed, labor, and husbandry.

In order to achieve aquaculture intensification in the rural areas, biotechnologies have to directed at low-input systems, marginal areas or to cater specifically to the rural community. However, all biotechnologies involve a significant development cost. It becomes difficult to recoup these investments making it inaccessible for most rural farmers. Moreover, application of these techniques has to be aided by scientific resources and capacity. Small hatchery operations increase the local supply of fingerlings and can enable farmers to enter aquaculture as an activity. They are integral to development of rural aquaculture but are hindered due to limited availability of water. As a result, they find it challenging to maintain the genetic quality of the stock being bred. In the long run, such stock may lose their performance and quality.

These situations can be remediated with the help of govern intervention or support from large hatcheries. The important factors to consider in these cases are the stage of rural development, the available extension programs and the ways by which these activities can be integrated with the existing approaches for supporting livelihood.

Another approach which is helpful in increasing the value of the local aquaculture farms is introduction of exotic species. For example, this practice has led to higher production of tilapia in Asia than in Africa which is actually the country of its origin. These exotic species are often genetically enhanced or domesticated to an extent. Hence, they face the similar risks and enjoy similar opportunities as the native species.

1.4. CHANGING PERSPECTIVES OF RURAL AQUA-CULTURE

The changing perspective of rural aquaculture are discussed in the following subsections.

1.4.1. Aqua Farmers Turning into Aqua Innovators

In order to maximize yields, farmers often experiment with the resources at disposal and the accessible technologies. The poor rural farmers cannot purchase expensive tools and techniques. With experience, the farmers come up with tried and tested and appropriate package of practices. This in turn is adopted by farming community carrying out activities in areas having similar agro-climatic conditions.

In many cases, these indigenous innovations remain unnoticed and the farmer is unable to completely utilize his talents. In order to promote rural aquaculture, it is important to nurture, promote and increase awareness about these innovations so that they can be used for planning and implementing research and development work. Local and national organizations often organize innovators meet to facilitate the documentation of such innovation and give recognition to the efforts of the farmers.

1.4.2. Ornamental Fish Village

Establishing ornamental fish village is an innovative technique of promoting rural aquaculture. For example, in India, Central Institute of Freshwater Aquaculture played an important role in developing two ornamental fish villages. Rearing ornamental fish in the backyard of the houses has become a popular trend and many cement tanks are being built for this purpose. This has emerged as a new way of generating income for rural women.

In many cases, the local agencies also provide financial assistance for helping set up these ventures. The farmers are often provided training for capacity building and are given exposure so that they can successfully carry on their ventures.

1.4.3. Gender Mainstreaming in Aquaculture

Involvement of rural women in aquaculture production activities including composite carp culture, seed rearing and integrated fish farming has been advocated for their socio-economic upliftment and generation of self-employment. One major challenge is that women have certain social and cultural constraints.

If aquaculture is extended using efficient techniques and effective technologies, it can encourage women to practice aquaculture in a sustainable manner. It is easy for women to manage backyard ponds and rear fish for a short period of time.

Even in most underdeveloped regions, aquaculture is emerging as a powerful tool to empower women. Various government initiatives and NGOs strive to facilitate the rural women to access Banks, Government establishments etc. They help women tackle the financial aspect of aquaculture management like purchasing input at a reasonable cost and accessing the market to sell the fishes. The rural women are also benefitted immensely from the additional income accruing from aquaculture, it improves their social status.

Figure 1.4: Rural women engaged in aquaculture.

Source: https://www.publicdomainpictures.net/pictures/30000/velka/women-operated-aquaculture-13446218490cp.jpg

1.4.4. Farmer Led Research to Gain Momentum

There is a general awareness that the existing extension systems are not sufficient to cater to the increasing needs of the farmers. The researchers have to deal with many challenges. The research system does not extend support to carry out demand driven research. As a result, there is a mismatch between need for research solutions at the farm level and its supply.

Generally, each country has its own research institute whose aim is to provide assistance to the farmers in their endeavors. They often launch project with the aim of bridging the disconnect and facilitates farmer led research. In the long run, it will provide valuable insights and help empower the farming community. The formal research system has to draw knowledge from the successful cases and improvise the innovations at the grassroot level.

There are several constraints that the farmers encounter when they innovate, these constraints have to be addressed and a solution has to be provided to the farmers. The farmer-scientist linkage has to be strengthened so as to encourage innovation, stakeholder participation, feedback and new institutional configuration. In other words, in matters related to aquaculture development, farmer participation in very important.

1.5. CONCLUSION

Aquaculture in rural areas is expanding rapidly all over the world. It is considered as an important source of protein, especially in emerging economies. It also forms and important part of farming systems and an important source of livelihood.

There is a decline observed in land and water resources. On the other hand, population is growing at an increasing rate, however, production from capture fisheries have become stagnant. Aquaculture has emerged as an important economic activity from national perspective. The expectations from aquaculture sector is high and will be targeted to increase in a sustainable manner.

There are sweeping changes taking place in many developing countries. However, in many regions aquaculture resources, still remain underutilized. Both horizontal and vertical expansion of aquaculture is important to achieve overall rural development. Rural aquaculture will continue to be important to the rural economy.

REFERENCES

1. Cifa.nic.in. (n.d.). *Rural Aquaculture* | Indian Council of Agricultural Research. [online] Available at: http://cifa.nic.in/node/328 [Accessed 29 August 2019].

2. De, H., & Pandey, D. (2014). *Rural aquaculture – Now and Then. Economic Affairs, 59*(4), p.497.

3. Edwards, D. (n.d.). *Rural Aquaculture*. [online] Cabi.org. Available at: https://www.cabi.org/bookshop/book/9780851995656 [Accessed 29 August 2019].

4. Fao.org. (n.d.). *FAO Fisheries & Aquaculture – Topics*. [online] Available at: http://www.fao.org/fishery/topic/16125/en [Accessed 29 August 2019].

5. Halwart, M., Funge-Smith, S., & Moehl, J. (n.d.). The Role of Aquaculture in Rural Development. [ebook] *Food and Agricultural Organization*. Available at: http://www.fao.org/3/y4490e/y4490e04.pdf [Accessed 29 August 2019].

2

Integration of Rural Aquaculture with Agriculture

CONTENTS

Figure 2.3: The aquaculture practices along with agriculture can provide the employment opportunities for the rural people.

Source: http://www.beachapedia.org/images/thumb/1/17/Aquaculture_shell-fish.jpg/300px-Aquaculture_shellfish.jpg

However, the industrial systems can be worked upon and enhanced to produce food for the world along with minimizing the negative impacts on the environment. These systems are also incapable to produce inexpensive food for the most of the people who lack security of food and various underprivileged others.

The only possible and sustainable way in which the underprivileged people can be provided fish food is by developing the small-scale or rural aquaculture. The systems like these are generally centered around the production of various species that are quite low in value and are meant for the normal consumption in the households or in the local markets.

The small-scale aquaculture systems are generally not considered worthy as their contributions are not reflected in the official statistics, but they make a major part in the aquaculture production across the globe.

The farming systems such as these are mainly used by the families that employ the farming practices of extensive or semi-intensive types for the rearing of various omnivorous or herbivorous species. The rural aquaculture normally forms a great part in the farming systems at broad levels, in which the other activities may be happening.

2.2. THE VARIOUS INTERACTIONS IN THE INTE-GRATED AGRI-AQUACULTURE SYSTEMS

The main purpose of integrating aquaculture into the farming systems that are already there, is to have maximum benefits form the best kind of interactions among the various activities. The systems in aquaculture, that belong to the regions that lack nutrients, need that the nutrients are supplied to increase the yields in significant ways.

These kinds of inputs are generally available in the form of the wastes that are generated from the animals and the crops, on the farms that are on small-scale. In the integrated systems of farming, the inputs are supplied majorly by the aquaculture in order to be supplied to other parts of the farming.

Figure 2.4: An initiative undertaken by some authorities from FAO focusing on the integration of aquaculture with agriculture.

Source: http://www.fao.org/typo3temp/pics/5ff45fcd53.jpg

The various byproducts in aquaculture that may involve the waste water in enriched forms and the sediments, can be employed in various activities in agriculture. There have been different kinds of methods which have been employed to analyze the various interactions that have been there among the components of the farms.

Figure 2.5: The rearing of fishes in the farms can provide the livestock with the required nutrients in the form of feed and can also have some benefits from them.

Source: https://www.agmrc.org/media/cms/aquaculture_7B47FD8474C2E.jpg

The methods employed treat all the interactions as a flow of nutrients, energy or mass, and hence there is always an undertaking of the budget analysis. The tools employed in the 1990s were mostly those of the bio-resource flow diagrams, in order to study the various interactions. However, these diagrams did not focus on the external flows that used to enter or leave the farms. These external flows were among the inorganic fertilizers or the products that were on the farms for sale. Thus, in the period following 1990s, there were other approaches, that were developed to analyze the interactions that considered all the flows. Mostly, the approaches developed later, employed a set of indicators and generally required the analysis to be done on software known as ECOPATH.

2.3. GENERAL CONSIDERATIONS IN THE INTEGRATION OF AQUACULTURE AND AGRICULTURE

The production of food eventually has some undesirable impacts such as:

- Negative environmental impacts
- The taking over and division of the natural habitats that primarily belonged to the various organisms
- The decrease in the biodiversity and the amount of wildlife
- Alterations in the quality of soil, water and the landscape

Most of the integrated agriculture-aquaculture systems have a smaller number of inputs and can be categorized into a kind of aquaculture known as semi-intensive aquaculture. This implied that they are not too much dependent on the large amounts of feed and other inputs in fertilized forms and even the lesser amounts of organisms that are farmed. Thus, they have lesser possibility of resulting in serious amount of pollution and lesser chances of giving rise to the risk of various diseases, in comparison to the increased amount of intensive systems that are feedlot-types.

This is an important observation and phenomenon because it is the large amounts of food products that are required in the feedlot-types systems of the intensive aquaculture that result in the increased levels of environmental pollution.

The systems that are of the semi-intensive kinds, interact with the agricultural systems, that forms an integrated kind of farming such as the one involving crop, livestock and fish, that makes use of the in-situ feeds, the proteins and the vitamins included, that form the feed for the natural aquatic forms. This also makes the need for the high-priced components of feed, necessary.

Figure 2.6: The integrated forms of aquaculture and agriculture in a field.

Source: https://hpdezign.com/wp-content/uploads/2012/12/
shutterstock_38490268-e1355366133768.jpg

The semi-intensive ponds of freshwater, generally have the environmental impacts in few forms that are different from the impact that they occupy the space of some of the previous natural habitats.

In the tropical regions, that result in the rapid generation of loading of organic waste, the wastes generated form these regions and the mud that is dug out, generally improves the productivity of the waterbodies and the land that lie adjacent to these areas and this avoids over nourishment of these areas.

In the place where there is a possibility of the construction of ponds and dike, making alterations in the subsoils of the acid sulfate and the places in which the alteration sin the water table may result in the upliftment of the salts on the subsurface, some special kind of care is important.

In addition, the intrusion of the saltwater that may come from the coastal ponds, may also contaminate the soils and the aquifers in freshwater bodies. The chemicals are generally not used in high quantities in the semi-intensive aquaculture systems.

However, there must be a great care that must be taken by the farmers while they are using the antibiotics or the hormones or some other types of drugs.

2.3.1. The Decision on the Use of Fish Species

The aquatic ecosystem may be shared by a lot of users and is capable of supporting a wide range of plants and organisms. With the development of the better quality of breeds in the domestic systems by the experts in aquaculture field, the demand for these breeds will keep on rising across the globe.

Figure 2.7: The Vietnamese have employed the integration of aquaculture with the practices of agriculture to provide themselves with economic gains and sustainable farming.

Source: https://ccafs.cgiar.org/sites/default/files/images/IMG_0021.JPG

This implies that the increased transfer of the premier breeds has been a great point that has benefited the farming related to crops and the livestock. However, the organisms that have gone through the rearing process and have grown up frequently tend to escape the breeding places and join the wild populations that can have certain consequences. These consequences may be given as:

- They may go to some other places and breed with other kind of wild animals or organisms, which may, in turn, lead to a situation of endangering the genetic resources of the natural ecosystem

- They may degrade the natural habitats of certain species as they can multiply themselves or result in the clearance of vegetation or may also result in an increased turbidity, which is also known as benthic forage

- They may result in the introduction of pathogens of the aquatic ecosystems, various such predators and similar kinds of pests, in an unintentional way.

The establishment of various agencies and the development of various farmers must be done by considering the benefits of using the exotic breeds and comparing them with the various environmental outcomes that may be possible from such use.

The projects and the farmers that are involved with the development, make use of the exotic breeds, without weighing the consequences and various implications, and only focusing on the benefits they provide.

Such an irresponsible behavior in the selection of breeds can result in unknown and unexplored consequences that may then be irreversible and may result in the loss or damage to the various habitats and the genetic resources that hold a lot of importance. This kind of damage may be there forever. The various norms that must be practiced in order to avoid such consequences have been developed in the recent times. However, the development of aquaculture still is not in the same frame as that of agriculture as far as the identification of the risks related to the transfers and the international usage of such breeds, and the measures to tackle them, are considered.

The guidelines that have, however, been laid down to be followed in aquaculture are:

- The farmers must use the native or local species and breeds which have been reared by the programs that have been conducted in local or national premises.

- If the need of the introduction of some other kind of breeds or species arises, the farmers must take some proper guidance from the experts in the aquaculture field so that they are capable of evaluating the probable consequences of rearing such breeds and have the knowledge on whether they are complying with the prescribed laws and the Codes of Practice, which have been laid down for the betterment and good of all the farmers, regardless of whether they belong to present times or the future.

2.3.2. The Consideration of Public Health

The integration of agriculture and aquaculture systems, normally, does not have any particular kind of health risk for the people, that is, even greater than the field of agriculture. However, the ponds containing the freshwater may be a cause for the spread of water borne diseases.

They may be the places that support the growth of parasitic worms, which may include bilharzia, and act as intermediate mediums for it and can also act as breeding places for the mosquitoes.

The problems like these are taken care of, by having ponds that are free of weed and stocked in a good way. Actually, the fish that reside in the ponds tend to consume a lot of larva of the mosquitoes and control it, but they are not capable of controlling the growth of snail in most of the cases.

The issues related to the accumulation of the pesticides in the fish reared in the rice fields are dropping in number as there has been an increase in the employment of programs that focus on integrated pest management, which tend to make more and more use of the natural substances and predators.

There is a very minute risk of the heavy metals that may come from the feeds of the livestock, accumulating in the sediments of the ponds for manuring and the fishes and is applicable to the intensive systems in quite an increased way.

In the similar manner, the pathways for aflatoxins, which are the poisons that are made from the fungi that grow in the feeds that are stored in poor ways, have had very little amounts of risk, but this area of study is still unexplored in many ways.

The culture of the fish in the sewage has been increasingly controversial topic as it is predicted that there would be a lot of risks to the health of the farm workers in such activities and also the fish that is reared in such capacities, having an implication on the health of the consumers, eventually. However, even these implications or risks may not be that high if they are compared to the benefits they provide to the consumers in terms of nutrients, only in cases where the fish, after their culturing or breeding, are managed in a hygienic manner and there must be an increased focus on not damaging the gut and hence, not permitting the contents of the gut to make contact with the flesh of the fishes. The products that come from such kinds of rearing and culturing also require to be cooked in a very thorough manner.

In the case of rearing of fishes in the sewage, there is not a presence of any common guidelines that focus on minimizing the risk on health from such rearing. In such cases, it is needed that the people associated with the process are aware of the various waterborne diseases that may be there in a certain area and assess themselves that whether it is sustainable to do such kinds of breeding as it involves the establishment of ponds and their proper operations and care, failing which it may lead to the health risks for the farm workers, the fish handlers and eventually, the consumers.

2.4. CONCLUSION

It is a common and a traditional practice to integrate various farming practices by using the methods of recycling, among quite a lot of societies in Asia, where such practices have taken the form originating form the process of 'slash and burn'.

The use of the aspects of ponds to have various uses in the farms and the other kinds of benefits they provide to the other components of a farm, such as the vegetable and the livestock, is quite a good feature of them. In such cases, the impacts that the semi-intensive ponds have on the environment, by the practice of conserving the habitats and species in the surroundings, may be positive in nature.

The activity of developing the various low-lying areas in the form of the ponds for persistent use, can help the various habitats that occur in its surrounding and may also be advantageous for the various organisms as it may be result in increasing the water that is available to these organisms.

One of the benefits that may be associated with the integration of the rearing of fishes with the other agricultural systems, is that it can help in controlling the spread of snails in the fields. This can be done by introducing some ducks in the fields that are watery and the various fishponds, and it can also help in controlling the spread of various other pests.

If the various points of views are considered, the various implications or the positive impacts of the integration of the agriculture with aquaculture and its practice and potential, must be analyzed properly. This must be done as the potential of the integration of agriculture and aquaculture may be quite high in the peri-urban regions. However, various factors such as

- the increased amounts of the pollutants and contaminants from the various household and the industries,

- the rising conflicts and competition for water and land and that too at the exorbitant prices,

- the rapid expansion of the cities which is intruding the agricultural lands, makes it quite difficult for the activities of such kinds to be successful.

REFERENCES

1.   Lightfoot, C. (1990). *Integration of Aquaculture and Agriculture: A Route to Sustainable Farming Systems*. [ebook] Available at: http://pubs.iclarm.net/Naga/na_2841.pdf [Accessed 29 August 2019].

2.   Prein, M., & Ahmed, M. (2000). Integration of Aquaculture into Smallholder Farming Systems for Improved Food Security and Household Nutrition. *Food and Nutrition Bulletin*, [online] *21*(4), 466–471. Available at: https://www.researchgate.net/publication/258993886_Integration_of_Aquaculture_into_Smallholder_Farming_Systems_for_Improved_Food_Security_and_Household_Nutrition [Accessed 29 August 2019].

3.   Pullin, R. (n.d.). *Integrated Agriculture-Aquaculture: A Primer*. [online] Fao.org. Available at: http://www.fao.org/3/Y1187E/y1187e07.htm [Accessed 29 August 2019].

4.   Shang, Y., & Costa-Pierce, B. (2009). Integrated Aquaculture-Agriculture Farming Systems: Some Economic Aspects. *Journal of the World Mariculture Society*, [online] *14*(1–4), 523–530. Available at: https://onlinelibrary.wiley.com/doi/pdf/10.1111/j.1749–7345.1983.tb00104.x [Accessed 29 August 2019].

5.   Van Huong, N., Cuong, T., Nang Thu, T., & Lebailly, P. (2017). Efficiency of Different Integrated Agriculture Aquaculture Systems in the Red River Delta of Vietnam. *Fisheries and Aquaculture Journal*, [online] 08(04). Available at: https://www.longdom.org/open-access/efficiency-of-different-integrated-agriculture-aquaculture-systems-in-thered-river-delta-of-vietnam-2150–3508–1000230.pdf [Accessed 29 August 2019].

6.   Zajdband, A. (2011). Genetics, Biofuels and Local Farming Systems. *Sustainable Agriculture Reviews*. [online] Available at: https://www.researchgate.net/publication/226829080_Integrated_Agri-Aquaculture_Systems [Accessed 29 August 2019].

3

Production Systems in Rural Aquaculture

CONTENTS

There have been several production systems that are being put to use in the rural aquaculture. But with the passage of time and advancement of technology, there have been many changes in the production systems that have affected the entire rural aquaculture. The main cause of the upgradation in the production systems was the rise of the human population and hence the increase in demand for the fishes and other products of aquaculture. There are several production systems that have been modified in order to meet the demands of the modern world.

3.1. INTRODUCTION

The sectors of rural aquaculture are considered to be on the rise. Especially in the growing countries like India, the economy had observed a great rise due to the integration of several production systems of aquaculture in the rural aquaculture.

Various production systems that are being used in the rural aquaculture have helped in recognizing the rural aquaculture as a powerful and significant source of income and employment. This is so because the rural aquaculture enhances the growth of several subsidiary industries. Along with this, rural aquaculture is a significant source of the nutritious food and a source of the earnings from the foreign exchange.

This can be very well understood by the numbers that have been recorded in the Indian Aquaculture market. In India, there has been a production of near about 10.79 million tons of fishes in the fiscal year 2016–17. With this kind of stats, India accounts for near about 5.68% of the fish production on the international platform. With the attainment of these numbers, India has reached the position of the second largest fish producer all over the world.

Not only in the production of fish, India has become the second largest producer of the aquaculture all over the world. The only country that has a standing above India in this market is China. As a matter of fact, India has been the home to more than 10% of the fish bio diversity that is present in the world.

It has been noticed that the global demand for fish has been on a rising trend in accordance with the increase of the human population in the world. This rise in the demand of fish is majorly due to the rich protein content. In this context, the production of fish and the practices of aquaculture have been gaining much importance. This importance has been given so that the compensation for the dipping numbers of the fish production from the main sources can be provided. Also, the rearing of the aquaculture can be considered as the primary substitute for the management of the sustainable fish production output.

In the recent times, rural aquaculture has turned out to be the largest increasing food industry of the world. From several studies and reports, it has been observed that there is an annual growth of near about 10% in the aquaculture as compared to the 2–3% increase that has been registered in other major food sectors.

Also, it has been noticed that rural aquaculture production has been growing at an average rate of near about 3.9% in the developed countries. When compared to the developing countries, the rural aquaculture has seen an average rise of near about 8.2%.

In several parts of the world, it has been observed that the farming and fishing activities are the primary livelihood of the people who reside in the rural areas. In order to boost the earning of the people and to boost their financial status, there have been several campaigns that are trying to join or integrate the aquaculture or the rearing of the fishes and other aquatic animals in ponds, with the conventional farming systems. This kind of integrations will be very helpful, particularly for the small-scale fisherman and locals who are practicing various production systems of aquaculture in the rural areas. It has been observed that the fish production and aquaculture in the smallholdings is usually restricted by the quantity and quality of the kind of input that is being provided to the pond. The main factor that puts the limit on the quality and quantity of the production of aquaculture is the timing of labor availability.

The unavailability of the required number of labors at the fish farm and several different kind of farm activities restricts the quantity of the inputs that the aqua culturists spend on the aquaculture. These kinds of restrictions or limits result in the lesser rates and decreased quality of the yields.

There is probability that the existing systems of production cane be improved, thus resulting in the increased production and better yields. These changes can be brought by making the required modification and changes in the production systems and the timing of the cultivation. As soon as these changes amendments are introduced in the production systems, there is a scope to accommodate other farming practices and activities.

Limited availability of the material and labor inputs in the various production systems of the rural aquaculture enterprises can be properly assigned by taking the seasonal availability of the required input materials into consideration and adapting the technology of the pond and fish farming as the major production system of the rural aquaculture.

Maintaining the focus on the technology that enhances the number of the fish production to the maximum extent is very important. This has become a more practiced system of aquaculture in the small-scale aquacultures as compared to the facilitation of implementation of the farming through the integrated fish farming.

For example, the Aquaculture subsector in the country of Zambia yields near about 10,000 tons of fish every year. Out of this 10,00 tons of fish, near about 75% of the entire yield comes from the several small-scale aquacultures. On the other hand, the major commercial fish producers contribute near about 25% of the entire yield.

3.2. PRODUCTION SYSTEMS AND SUSTAINABLE FOOD PRODUCTION

There are several studies and reports that show that the usage of the aquaculture for the rural development does not have an impressive record in most of the developing countries. There are some other review studies and reports which show that aquaculture does help the poor to earn a better livelihood, especially in the areas of Asia where the rural aquaculture is a conventional practice. But there are several constraints that obstruct the expansion and better establishment of the rural aquaculture in these areas.

The adoption of modern technology in the recent times have suggested that, with the correct support, rural aquaculture can also provide help, in a significant number, to the area of rural development in those countries where it is neither a conventional practice nor a general practice.

Rural aquaculture in the developing countries can help in enhancing the sustainability of all the small-scale aquacultures. But this can be only possible when the rural aquaculture is completely integrated with several other farming practices and household activities. When the complete integration of the rural aquaculture is done, it allows the small farm families and communities of rural aquaculture to manage their natural resources in an efficient manner.

Figure 3.2: Different production systems are essential for the establishment of the sustainable practices that could help in the betterment of the rural aquaculture.

Source: https://upload.wikimedia.org/wikipedia/commons/b/b5/Seaweed_ farming_-Nusa_Lembongan%2C_Bali-16Aug2009_edit.jpg

The rural aquaculture is a weapon to fight against poverty and decrease the extent of inequality has attained renewed attention. In addition to it, it contributes to poverty alleviation and offers the employment services to millions of the individuals, both in the sector itself as well as in support services.

3.3. DIFFERENT TYPES OF PRODUCTION SYSTEMS

3.3.1. Freshwater Production Systems

Freshwater production system in rural aquaculture is a significant component of the supply of animal-based protein, amino acids, fatty acids, minerals and vitamins in the human food consumption.

The freshwater production systems were anticipated to turn out to be more exhaustive and uniform. It was also predicted that the output of freshwater production systems will, on the one hand, fulfill the ever-increasing demand of the common marketplaces for the safe and secure animal-source products.

3.3.1.1. Pond Fish Culture

It has been noted that at many places, such as in China, the freshwater production system in rural aquaculture primarily refers to the pond fish culture. In the year 1997, it was observed that the pond fish culture area acquired near about 2 million hectares, which did not comprise of the paddy field culture area, with an output of near about 8.9 million metric tons. This huge output accounted for near about 72.1% of the entire freshwater production in the rural aquaculture.

The technological system that is being out to use in the pond fish culture is chiefly the conventional Chinese fish farming technology, which was being used by the human beings in the early ages. In the modern times, this technology is just refined and enhanced with the help of the knowledge and understanding that has been gained from the years of research and several attempts that have been made for the development of the pond fish culture. The pond fish culture of the production system of rural aquaculture has the following outstanding features:

Figure 3.3: Pond fish culture focuses on managing the water quality along with [promoting integrated aquaculture.

Source: https://www.nps.gov/puho/learn/historyculture/images/2010525_ PUHO_156_1.jpg?maxwidth=650&autorotate=false

- Cultivating short food-chain fish: The fish that is cultivated or reared in the region of China is mainly herbivorous or omnivorous. Thus, these kinds of fishes have a very short and compact food chain. Fertilizer, grass, waste products that are generated from the farm products processing industry can be majorly utilized as the food for the fish.

So, the sources of their feeds are ample and widespread, and the cost of rearing or cultivating these kinds of fishes is quite low. This low cost usually results in the improved financial benefits.

- Self-sufficiency in the production of seed: in the present times, it has been noticed that more than 20 species are being reared or cultivated with the help of artificial methods. The seeds that are produced in the hatcheries, belong to all the primarily culturable species of fishes. These seeds are usually available in near about all the areas where the rural aquaculture is a significant practice.

Therefore, it is recommended that rural aquaculture activities must be done in a pre-planned way and must be done in accordance with the demands of the fishes and other products of the aquaculture. This helps in the proper and sustainable usage of the resources.

- Integrated practices in rural aquaculture: The practice of pond fish culture is usually done with integrating it with a number of other farming activities. Some of the activities that are usually integrated with the pond fish culture are the rearing or cultivation of the livestock. Some of the major animals that are reared along with the rural aquaculture are the chicken, duck, pig, and cattle. Along with livestock rearing, the practice of crop cultivation or horticulture is also done.

Cultivation of vegetables, mulberry, fruits, and ornamental plants is also done along with the rural aquaculture. These kinds of inclusive and integrated methods of production along with the rural aquaculture have several benefits. While keeping the aquaculture as the primary practice and cultivating a variety of cash crops, grass as feed, and rearing of the livestock and poultry near about the site of aquaculture can help in strengthening the economy of the rural areas.

The waste of reared animals can be fermented and utilized as fertilizer or, in some cases, as fish feeds. Similarly, the sludge or the mud that is present at the bottom of the pond can be utilized as the good-quality fertilizer for the crops that are being cultivated on the land. Also, the crops and grass that are being cultivated can be utilized as the food for the animals on the farms and also can be used as the feed for the fish. This kind of integration permits the best possible uses of all available resources and thus results in generating a higher household income for the small -scale farmer in the rural aquaculture.

- Water quality management: For the proper and healthy cultivation of the fishes, it is very important to maintain the quality of the pond water. If the quality of the pond water is maintained at the optimal level by maintaining the balance in the pond ecosystem, it results in the better quality of the fishes cultivated in the rural aquaculture. This can be done with the help of careful management of the feeding regime, inflow and outflow of the water, and the periodic aeration of the water body.

3.3.2. Cage Culture

Figure 3.4: Cage culture is a type of production system in rural aquaculture that focuses on customized development of the fish breeding.

Source: https://upload.wikimedia.org/wikipedia/commons/8/87/Rupa_lake_cage_culture.jpg

The development of the cage culture took place in China. Chinese modern net cage culture started in the early days of the year 1970s and the main attempts were focused on developing and extending the method in the year 1980s.

In the present times, net cage culture is being used in lakes, reservoirs, rivers, ditches and shallow seawater and can be largely segmented into four types:

- To make use of the natural feed (planktons) to culture large-sized silver carp and big-head carp fingerlings for the stocking up the large water bodies;

- To make use of the natural feed (planktons) to culture table fish;

- To make use of the artificial feed to culture table fish; and

- To culture high value species like eel, mandarin fish, sea bream, and many more, through high density intensive feeding system. Net cage culture in the developed nations is generally an intensive culture system, but the culture system in the region of China is either traditional or it is semi-intensive.

3.3.3. Paddy-Fish Culture

Paddy-field fish culture in the region of China has a history for over two thousand years. In the initial days of 1970s, China carried out widespread investigation on ecology and biology of culturing fish in rice-fields.

This gave rise to the development of several methods of rice-fish culture, based on a symbiotic connection among fish and rice cultivation, leading to enhanced economic profits to the farmer. The main kinds of rice-fish culture are as follows:

- Raise fingerlings in paddy fields in the plains: The fry is directly reared or cultivated in early rice fields. After the fry have grown to four to five cm, they can be relocated to the semi-late rice fields to grow them to market size by the time the rice is harvested. This is the considered as the easiest and most efficient way which brings the maximum profits to fishermen.

- Planting rice on a ridge and rearing fish in a ditch: The best specifications are a ridge, comprising of two rows of rice seedlings and a ditch of one meter in width and one meter in depth. Some of the farmers even plant melons and soy-beans on banks and rear plants like duckweeds in the water body. And therefor, a kind of multi-layer planting and culture mode is formed.

- Rearing fish in a wide ditch: Broader ditches of difference sizes are created in the rice-field and are associated with the water inlet for fish culture. The entire or total area for fish culture may differ from 5–10% of the rice-field.

- The other method of rice-fish culture is to cultivate rice and fish in rotation in the similar field.

The above stated culture methods are wide, however in some of the cases supplemented feed are given to the fish.

3.3.4. Maricultural Production Systems

Before the year 1950s all maricultural systems in the nations were extensive kind. Fish and shrimp larvae were used to be trapped at the time of high tides in ponds with sluice gates. These sluice gates were created on the mudflats, and several kinds of substrates were used in order to settle the spat.

The effective research and developmental activities of the previous four decades resulted in the enhancements as well as the modernization of culture systems. In the present times, most oyster, mussel, shrimp as well as seaweed culture are of semi-intensive kind.

This has been made possible by the success in artificial method of breeding of many marine species for rearing seed for the purpose of stocking, in the development of formulated feed, in the health management of cultured organisms, and many more.

Generally, mariculture in the region of China is segmented into mudflat culture and shallow sea water culture. Mudflat culture is defined as the culture which makes use of intertidal area. It has had a long history and places emphasize on culture of oyster, constricted tagelus and bloody clam.

The methods of Mudflat culture are as follows:

- planting seeds of shellfish in mudflat directly, like constricted tagelus and many more;

- making use of the stone, bamboo pole and many more, as an attachment to gather the seeds and spats like oyster, gracilaria, and cultured in mudflat;

- building of pond in mudflat for prawn and fish culture. In the year 1997, the mudflat culture output made up 50% of the entire output of Mariculture.

Mainly, the shallow sea culture is employed in the places which are close to gulfs and islands, where the flow of water is smooth; quality of water is good and there are shelters against stormy waves. By the use of racks, floating rafts, floating ropes, and many more, kelp, laver, gracilaria, undaria, oyster and mussel are cultured. Fish culture is also carried out with the help of net cages.

3.4. PRODUCTION FACILITIES OF THE MODERN WORLD

Rural aquaculture facilities in the region of China are very simple. These facilities are simple in those regions where the farmers give importance or priority to agriculture and aquaculture is only considered as their sideline production.

For instance, in paddy field fish culture the most basic facilities are paddy fields and their linked ditches for water to flow in and out, sluice gates; on the other hand, in Mariculture are earth dykes and bamboo wooden sticks, and many more.

For water supply and draining of ponds, different kinds of general-purpose water pumps are mainly used. In some small districts where pond culture is not very well developed, water supply and drainage are carried out by small pumps for individual pond. On the other hand, if pond culture is well developed with clusters of ponds, there is a use of pump station in order to pump and convey water through distribution pipes to the ponds nearby. Heavy machineries like bulldozers and dredgers are used for the purpose of excavation/desilting. Also used are locally made aerators of various kinds such as impeller, paddle wheel, sprinkling and jet flow type are used to supply oxygen to the pond waters.

The fish farmers also make use of the locally made simple machines for the purpose of crushing snails and clams, green feed cutting machine is used for cutting aquatic plants into small pieces or as a palp for feeding fish or it can also be used as a pond water fertilizer. Also, the machines for making pellet feed in the small fish farms are used. For the purpose of harvesting, winch and rope winding pulley are generally used for the purpose of drawing the harvesting net from one end of the pool to the other end of the pond.

The ancient or old structures of production like the production team, production brigade and people's commune vanished and thus the large production units no longer occur. With the promotion of production-related contract responsibility system, the rural aquaculture is essentially controlled by households. In most of the scenarios, every single household contract a pond which usually covers near about one hectare. Small-scale lakes and reservoirs are mostly contracted by various households.

10. Fao.org. (2019). *Systems Framework for Rural Aquaculture*. [online] Available at: http://www.fao.org/3/x6941e/x6941e07.htm [Accessed 29 August 2019].

11. Fisheries and Aquaculture. (2018). [ebook] Mumbai: Farm Sector Policy Department NABARD Head Office, Mumbai. Available at: https://www.nabard.org/auth/writereaddata/tender/0803193019Fisheries%20and%20Aquaculture.pdf [Accessed 29 August 2019].

12. Lightfoot, C., Bimbao, M., Dalsgaard, J., & Pullin, R. (1993). Aquaculture and Sustainability through Integrated Resources Management. *Outlook on Agriculture*, [online] *22*(3), 143–150. Available at: https://journals.sagepub.com/doi/abs/10.1177/003072709302200303 [Accessed 29 August 2019].

13. Nal.usda.gov. (2019). *Aquaculture | Alternative Farming Systems Information Center | NAL | USDA*. [online] Available at: https://www.nal.usda.gov/afsic/aquaculture [Accessed 29 August 2019].

14. Prein, M. (2002). Integration of aquaculture into crop–animal systems in Asia. *Agricultural Systems*, [online] *71*(1–2), 127–146. Available at: https://www.sciencedirect.com/science/article/pii/S0308521X01000403 [Accessed 29 August 2019].

15. Rahman, A., Uddin, M., Chowdhury, M., & Shefat, S. (2018). Strength, Weakness, Opportunities and Threat Analysis of Integrated Aqua-farming in Bangladesh. *Acta Scientific Agriculture*, [online] *2*(12), 112–118. Available at: https://actascientific.com/ASAG/pdf/ASAG-02-0269.pdf [Accessed 29 August 2019].

16. Rajee, O., & Kar Mun, A. (2017). Impact of aquaculture on the livelihoods and food security of rural communities. *International Journal of Fisheries and Aquatic Studies*, [online] 5(2), 278–283. Available at: http://www.fisheriesjournal.com/archives/2017/vol5issue2/PartD/5-1-81–919.pdf [Accessed 29 August 2019].

17. Review of the State of World Aquaculture. (2003). [ebook] Fishery Resources Division FAO Fisheries Department. Available at: http://www.fao.org/tempref/docrep/fao/005/Y4490e/y4490e.pdf#page=50 [Accessed 29 August 2019].

18. Sarker, S., Basak, S., Hasan, J., Hossain, M., Rahman, M., & Ahsanul Islam, M. (2017). Production in Small Scale Aquaculture Farm: A Success Story from Bangladesh. *Journal of Aquaculture Research & Development*, [online] *8*(10). Available at: https://www.longdom.org/open-access/production-in-small-scale-aquaculture-farm-a-success-

story-from-bangladesh-2155–9546–1000506.pdf [Accessed 29 August 2019].

19. Van Huong, N., Cuong, T., Nang Thu, T., & Lebailly, P. (2017). Efficiency of Different Integrated Agriculture Aquaculture Systems in the Red River Delta of Vietnam. *Fisheries and Aquaculture Journal*, [online] *8*(4). Available at: https://www.longdom.org/open-access/efficiency-of-different-integrated-agriculture-aquaculture-systems-in-thered-river-delta-of-vietnam-2150–3508–1000230.pdf [Accessed 29 August 2019].

20. Zira, J., Ali, M., Ja'afaru, A., Badejo, B., & Ghumdia, A. (2015). Integrated fish farming and poverty alleviation/hunger eradication in Nigeria. *IOSR Journal of Agriculture and Veterinary Science*, *8*(6), 15–20.

4

Aquaculture Technologies: Concepts and Approaches

CONTENTS

4

Aquaculture Technologies: Concepts and Approaches

CONTENTS

Aquaculture technologies are used in the aquacultural sector so as to get maximum improvements to the culture. This chapter is all about the aquaculture technologies including the concepts and approaches. There are some of the aquaculture technologies such as well-boat technology, recirculating agriculture technology, nanotechnology in aquaculture, farm management software, smart floating farms, aquapods, seabass hatchery, UV systems and electro pulse fishing which are described in this chapter.

4.1. INTRODUCTION

In last few decades, the technologies and the systems which are used in the aquaculture have been developed rapidly. In this, a variation can be seen from very simple facilities to the high technology systems. Most of the technologies used in the aquaculture are relatively simple and are mostly based on the small modifications and upgradations which further improves the growth and survival rates of the species targeted.

Improvement in food, seeds, and oxygen levels, among others. Simple systems which are used in the small freshwater ponds are used for the purpose of raising herbivores and filter feeding fishes which is responsible for around half of the global aquaculture production.

Figure 4.1: Aquaculture tanks.

Source: https://www.flickr.com/photos/bytemarks/5211291608

The development of the closed systems has taken place by having a greater understanding about the complex interactions in between the nutrients, bacteria and cultured organisms along with the advances and modifications in the hydrodynamics which are further applied to the ponds and tank design. There are many advantages of isolating the aquaculture systems from the natural aquatic systems which further minimizes the risk of the disease or genetic effects on the external systems.

The advancements and developments of the aquaculture and related technologies have taken place in last few decades but still there are some of the technologies that have not been adopted readily by farmers. There are critical factors which influence adoption decisions in the aquaculture technology such as:

- method of information transfer.
- characteristics of the technology.
- farm characteristics.
- economic factors.
- sociodemographic and institutional factors.

Fish farmers are adopting some of the technologies which are much advantageous that other in terms of the cost efficiency, ease of management and also production. There are key economic factors which influence the adoption decisions such as the price of the aquaculture products and the profit expectations from the business ventures.

4.2. AQUACULTURE TECHNOLOGIES

4.2.1. Well-Boat Technology

Well boats are the unique kind of fishing ad housing facility vessels which provide the invaluable assistance to the whole fisheries sector both in terms of the commercial viability and the end-consumer's satisfaction.

The concept of using the well boats as the fishing boats initiates from a purpose which is very much focused in order to provide the qualitative fisheries. There is a traditional aquatic culture approach that involves catching of fishes and further results into preserving the fishes in order to retain its usability.

It is the latter aspect that acts as the main driving force responsible for the immense success of the well boat worldwide. There is no threat of the contamination of the fishes and other complications which is possible because of negligence in the storage since, they can thrive until the last stage of their processing.

There are some facts related to the well boats such as:

- The designing as well as storage aspects of each fishing boat is unique and bears distinctive features singular to its own.

- Though the storing capacity differs, an ideal boat can have storing capacity of about 1,400 cubic meters and can house up to 180 tons of fishes.

- In order to transfer the fishes into the boat, connections are built between the huge pipelines and the fishes' breeding enclosures in the high seas.

- These pipes help in transferring the fishes into the boat's container area without disturbing the fishes.

4.2.2. Recirculating Aquaculture Technology

There are intensive fish production systems which can be described by the Recirculating Aquaculture Systems (RAS). These intensive fish production systems can use a series of the water treatment steps in order to cleanse the fish-rearing water and further facilitate its reuse.

Recirculating Aquaculture Systems can generally be included in:

- Devices in order to remove solid particles from the water which are composed of fish faces, uneaten feed and also bacterial flocs,

- Nitrifying biofilters so as to oxidize ammonia which is excreted by fish to nitrate and

- A number of gas exchange devices for removal of dissolved carbon dioxide which is expelled by the fish and/or adding oxygen required by the fish and further nitrifying bacteria.

Furthermore, the recirculating aquaculture systems can also use UV irradiation for various purposes such as water disinfection, ozonation and protein skimming for the fine solids and microbial control as well as the denitrification systems so as to remove the nitrate.

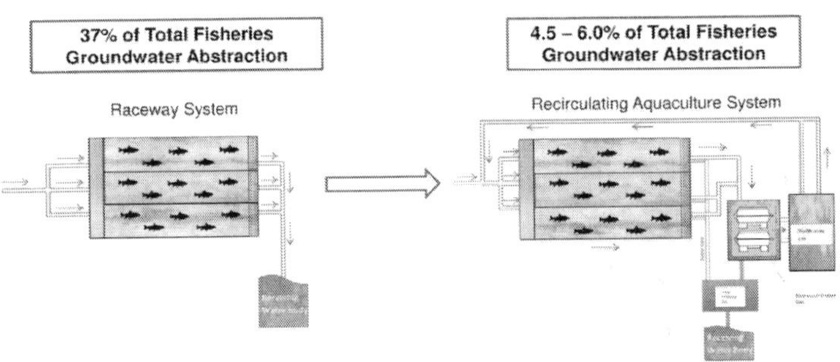

Figure 4.2: Switching from a regular fish farm to recirculating aquaculture systems.

Source: https://commons.wikimedia.org/wiki/File:Switch_from_a_Regular_fish_farm_to_RAS.jpg

There have been developments in the modern recirculating aquaculture technology for more than 40 years. Despite this fact, the new technologies increasingly offer a number of various ways in order to change the paradigms of traditional recirculating aquaculture systems which include improvements on classic processes like solid captures, biofiltration and gas exchange. These systems have also seen some of the important developments with respect to scale, production capacities and market acceptance and this has made the systems progressively larger and more robust.

There are different solutions in aquaculture which have been designed in order to increases the water usage including the recirculating systems which incorporate the water filtration systems like drum filters, biological filters, protein skimmers and systems related to oxygen injection.

Figure 4.3: Recirculating aquaculture system.

Source: https://commons.wikimedia.org/wiki/File:Recirculating_Aquaculture_System_7.jpg

The trust in the technology was regained because of the successful operation of the public as well as the domestic aquaria that generally contains over-sized treatment units in order to ensure the crystal-clear water. In addition to this, there is extremely low stocking densities and related feed inputs which over-engineering still made a small contribution to capital and operational costs of the system in comparison to the intensive recirculating aquaculture systems.

4.2.3. Nanotechnology in Aquaculture

In aquaculture, the nanotechnology can be used in variety of ways and in variety of spectrums. This technology can significantly contribute to its evolution. This technology can be used in many ways such as in the sterilization of the ponds, water treatment, detection and control of the aquatic diseases, efficient delivery of the nutrients and drugs which includes hormones and vaccines and also in the improvement in the ability of fishes so as to absorb the substances.

There are certain parameters such as physical, chemical and biological that should be considered for the quality of water and this can further interact individually or collectively by influencing the aquaculture productivity. The loss in the quality of water can lead to the poor performance and injury to the organisms.

The use of nanotechnology for the purpose of the water treatment has gained ground worldwide. This offers a number of applications which can be specific to the customers. In these systems, there are many useful aspects such as nanoadsorbents, nanometals, nanomembranes, and nanophoto-catalysts. These aspects are the ability to integrate the variety of properties in multifunctional materials.

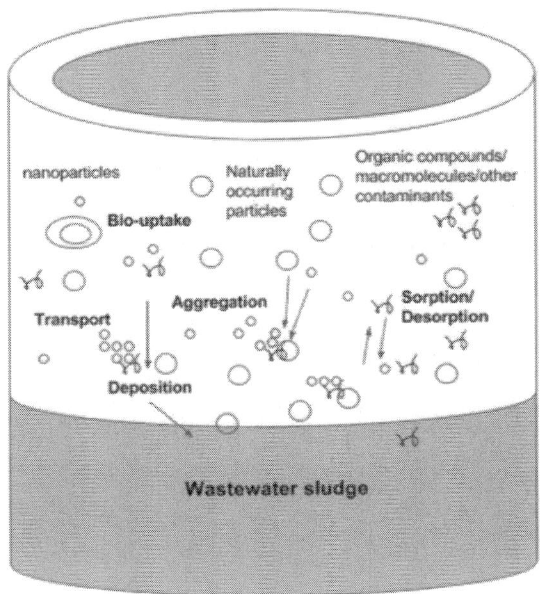

Figure 4.4: Silver nanoparticle interactions in wastewater treatment systems.

Source: https://commons.wikimedia.org/wiki/File:Silver_Nanoparticle_Interactions_in_Wastewater_Treatment_Systems.png

Let's take an example, nanomaterials can be used for the simultaneous removal of the particles and the removal of the contaminants, further providing the greater process efficiency. There are many losses which are faced by the aquaculture industry in annual basis because of the diseases which are caused by various pathogens. And so, the efficient detection and control of the diseases is very important for increasing the productivity and so as to ensure the satisfactory quality of the final product. The nanobiosensors can be further used and provide various innovative ways so as to solve the problems. There are devices which can be further based on the various nanomaterials such as carbon nanotubes, which makes it possible in order to detect the low concentrations of the pathogens which includes the bacteria, viruses, parasites and also pollutants.

Along with the detection of the aquatic diseases, nanotechnology can also provide its contribution in order to improve their control. So as to solve the problems related to the traditional methods of disinfection and sterilization like environmental contamination leading to excessive use of chemicals, more efficient formulations needs to be used.

The researches which are conducted in this industrial sector, the major focus was kept on the development of the nanotechnological formulation for the application in aquaculture. These systems contain many features like they are suitable for various application such as administration of vaccines, antibiotics, other pharmaceuticals and nutraceuticals.

Some of the advantages include:

- improved bioavailability of active agents that present low absorption,
- sustained release of the active agent,
- decreased frequency of application,
- high molecular level dispersion, due to nanoscale size, which can be used in the treatment of genetic diseases.

Conventionally, the application of active substances is achieved in the food or by direct injections in fish. This can be further related to the substantial losses during the processes along with the side effects occurred because of the excessive use.

Nanotechnology can contribute to the production as well as the marketing of the fishery products, especially when there is a widespread need of the extension production shelf-life. Therefore, this have intensified the research efforts for the food packaging sector. Because of the increasing environmental concerns, there has been an increase in the usage of the biopolymers in the food industry. However, there are still many problems faced by the sector which is related to the high production costs and the performance of these materials in comparison to the synthetic polymers. The use of the nanomaterials in the biodegradable food packaging can further increase the performance by improving the mechanical as well as thermal properties of the materials. In addition to this, there are new properties provided by the nanomaterials like antimicrobial and antifungal activities, enzyme immobilization, protection against degradation and elimination of oxygen, further contributing to the better stability of the stored products.

Despite their potential, the representation of a source of new contaminants released to the environment and the possible negative effects that can occur because of the exposure of these substances, is done by the nanomaterials. Nanotoxicology is the field of science which studies the major effects of these materials on the different organisms.

In such way, it can be said that the use of the nanomaterials and the nanoformulations which are based on the natural polymers and matrices can be an alternative so as to solve the problems related to the fishing industry and the aquaculture. This can provide its contribution in the reduction of the contamination and further providing more safety to the consumers and the producers.

4.2.4. iFarm System

There are sensors in the iFarm systems that have the computer vision. This helps in recognizing each individual which is based on the dot pattern of the salmon. In the chamber of the sensors, the number of fishes, the fish size, the number of sea lice and some of the possible signs of the disease gets registered. This kind of method allows the individual-based fish farming. iFarm has provided a technological leap in the development of the cage farming.

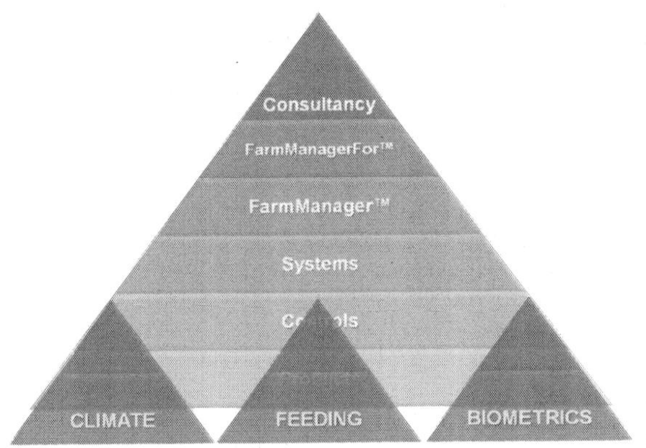

Figure 4.5: iFarm pyramid.

Source: https://commons.wikimedia.org/wiki/File:IFarm_Pyramid.jpg

The growth in the channeled way to the sites of the current type can provide huge profits. This strategy for the purpose of growth will preserve as well as strengthen. iFarm will give a significant contribution in the solution of the area challenges which are faced by the aquaculture.

When the fish passes the sensor chamber, the individual fish can be taken out for example the lice treatment, since the lice are very unevenly distributed. So, the lice treatment can further be reduced if we only treat the fishes that contain lice. In the similar way, the sorting is done on the basis of weight and removes the fish which is ready for harvest without stressing the fishes remained in the chamber.

Monitoring of each individual fish is done in order to detect if any fish has stopped growing and so it differentiates from the previous growth rates or it can get some unexplained decrease in the condition factors. These can act as the symptoms that there is something which is not right in this fish. Detection of the changes even if the observation is within the normal of the population is done and this is possible because the iFarm has an entire growth and conditional history. The diagnostic work and measures can further be initiated earlier.

iFarm has proved to be a technological leap for the cage-based salmon farming where shifting from the group-based operations takes place to the individual registration and treatment. There are iFarm sensors which is a technical concept developed by the BioSort AS. BioSort's key personnel have all background from Tomra's recognition and sorting technology. A number of new opportunities can be opened up by further development in these technologies to be applied in the sea.

4.2.5. Smart Floating Farms

The floating farm is an offshore structure which can be built at the coast of a city so as to produce the fish as well as vegetal food. This can be done throughout the simple system of connections in between different operative layers. These types of floating farms can easily become an automated farm clusters which are run by using the information and technologies and further software.

These processes can be managed automatically or with small interference done by the human. For the report data, the sensors can be placed everywhere within the structure which makes the structure rely on the Internet of Things. Ultimately, this makes the entire structure to run on remote and will not require any physical presence.

Figure 4.6: Floating fish farm.

Source: https://www.flickr.com/photos/terrazzo/36350079763

Furthermore, through the application of the modern big data management's tools it would be possible to determine what the local populace need most from the floating farms.

4.2.6. Aqua Pods

A unique containment system for marine aquaculture is known as the aqua pods which are suited for open ocean conditions and also a diversity of species. There are individual triangle panels in the aqua pod which are fastened together in a spheroid shape.

The aqua pod can also be referred to as a cage in which the farmers can keep their fish and then leave it floating in the sea. One small aqua pod is attached to a manned sailing vessel in the deep ocean. The engines in the vessels are used occasionally but is only used in order to correct the course of the floating fish farm and so helps in minimizing the amount of diesel which is to be used.

4.2.7. Seabass Hatchery

Seabass can be created easily by using a hormone and its larvae which are reared in the hatchery with 90 % survival. The only possible problem for the hatchery operator is that the seabass are protandrous hermaphrodites. This means that they first mature as males then become female in their sixth year or even this can happen when their size is over 3 kg.

Should egg production decrease as a result of having more female brood stock, the operator must cull and acquire younger or male brood stock.

Wild spawners can be used as breeders to be raised in cages, tanks or even ponds. Seabass are injected with the help of LHRHa that is Lutenizing Hormone-Releasing Hormone. Then they are left to spawn for 2 to 3 consecutive days. Live foods such as Brachionus and Artemia, are mostly fed to the larvae which was hatched from the eggs.

4.2.8. Ultraviolet System in Aquaculture

There are ultraviolet systems which are used in the aquaculture such as ULTRAAQUA UV system. This type of system disinfects more than 100,000 m^3/h of water in aquaculture systems in a continuous manner and millions of sturgeons, turbot, seabass, salmon and eels etc. are produced in the aquaculture system across the world.

In this scenario, the Ultra aqua UV systems are chosen in order to increase the security from the infection diseases by further protecting millions of invested dollars.

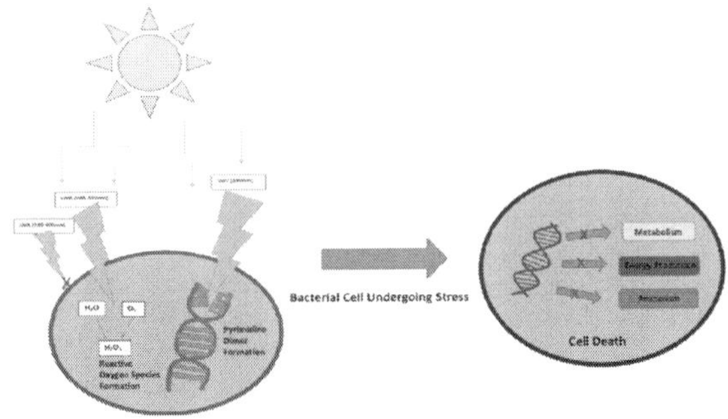

Figure 4.7: UV radiation to kill bacteria.

Source: https://commons.wikimedia.org/wiki/File:Figure_1_-_UV_radiation.png

Ultra-aqua UV systems are used in Chile so as to prevent from the diseases such as Infectious Salmon Anemia (ISA). This has provided the fish farms, much security and reassurance that the fish is not infected. The UV systems are also very easy to maintain and have the lamp lifetime of 16,000 hours and they also do not take up time in the daily routines.

This is reason, why the UV systems are highly recommended and ae used in most of the aquaculture systems in the entire world. With the increase in the disease concerns which are found in the source waters, the water abundance and the purity continues to decrease.

Concurrently, with the increasing consumption of fish, the demands for the higher stock densities in the same hatchery footprint are also increasing. Aquafina UV Systems are perfect to be used in the fish hatcheries, incubation, rehabilitation facilities, depuration facilities, aquariums processing plants influent treatment.

4.2.9. Electro Pulse Fishing

A technique in which the electrical energy is put into the water and fish is known as electrofishing. The technique of electrofishing depends upon the two electrodes which provides direct current a high-voltage from the anode to the cathode through the water. When a fish discovers the large enough potential gradient on this path, it gets affected by the electricity.

Usually, the pulsed direct current (DC) is applied which can cause the muscular vibration in the fish. The fish intercepts the energy and so are drawn towards the probes and incapacitated in such a way that they can further be captured with the nets.

The movement of the fishes towards the source of the electricity is known as galvanotaxis. This is the uncontrolled involuntary muscular convulsion which can lead to the fish swimming towards the anode. It is believed that this is a result of direct stimulation of the central and autonomic nervous systems which tend to control the voluntary as well as involuntary reactions of the fishes.

4.3. APPROACHES TO THE AQUACULTURE TECH-NOLOGIES

4.3.1. Technological Approach

With the continuing worldwide exploitation rate of capturing fisheries, the development of the sustainable aquaculture practices is increasing in order to meet the seafood needs of the growing world population. The demand for aquatic products was historically satisfied firstly by an effort to expand wild catch and secondly by increasing land-based and near-shore aquaculture.

Furthermore, the stagnation in the wild catches along with the environmental and societal challenges of the land-based and near-shore aquaculture has much promoted the efforts to the development of farming offshore technologies for harsh, high energetic environments. As a result, ocean domestication seems to be very important in order to maintain the oceans as a sustainable source of food both economically as well as ecologically.

While predicting the biomass, the sensors and techniques are still insufficient. But there are some other devices such as submersible frames which count and estimates the size of the fish passing through it with the help of optical techniques such as biomass counters or biomass estimators. Automated systems are used which uses the computer vision in order to monitor the fish passage through the special devices in the rivers. This also needs to be considered that most of the fish species are very reluctant to pass through the artificial devices which further increases their level of stress. The measurements of the fish passing through these devices cannot represent the whole fish population in the cage. In last few years, the techniques have been developed that uses the underwater video monitoring that have been proposed by a number of several authors in the aquaculture domain.

Despite this fact, there are still many systems which are still operated with the low level of the automation and further require intense activity from the operators. The computation of 3D variables is required by the fish segmentation and measurement. This computation includes the distance from the fish observed, fish orientation and size.

So, the implementation of the 3D computer vision should be done. However, the cost of the systems using 3D computing visions are very high for many fish farms and cannot be possible applied because of the difficulties in the fish farm environment. Moreover, the hardware and software computational requirements of the system have been minimized by avoiding complex 3-D fish models.

This technological approach indicates that there are some systems which use simplified models which are based in the fish features and existing relationship between the fish length and its weight. This approach has provided considerable advantages for its simplicity, modularity and flexibility in order to adapt different fish species. Despite all this, it requires lower computer requirements in comparison to other segmentation methods which are based on the 3D models or active contours.

4.3.2. Systematic Approach

The system approach is a multifactorial and multidisciplinary approach and furthermore, it recognizes the diversity of the influences on the aquaculture development. This kind of approach uses an understanding about the operation of the aquacultural systems in order to analyze the different factors that affect the aquaculture and also develop the solutions to the problems which are identified.

This kind of analytical approach has provided its contribution in order to identify the key researchable issues, develop the better management solutions, improve the business efficiency and design and test the new aquaculture systems along with the more effective extension and education.

The systematic analysis and understanding of aquaculture systems can occur at a number of different levels:

- the organism and its surroundings,
- the production unit,

- the economic enterprise,
- the farm, watershed or coastal areas,
- the national sectoral level,
- the international level.

The boundaries which are chosen for the system of interest can be either physical entities, organizational structures or the political borders. This systematic approach recognizes the fact that there are important interactions in between these different levels and further majorly focuses on the following systems:

- the farm
- the local environment
- the national level

REFERENCES

1. Aquaculture technology. (n.d.). [ebook] p.1. Available at: http://www. fao.org/figis/pdf/fishery/technology/aquaculture/en?title=FAO%20 Fisheries%20%26%20Aquaculture%20-%20Aquaculture%20 technology [Accessed 29 August 2019].

2. Cermaq.com. (2016). *Cermaq | iFarm-Cermaq Towards Individual-Based Farming*. [online] Available at: https://www.cermaq.com/wps/ wcm/connect/cermaq/news/ifarm-cermaq-towards-individual-based-farming [Accessed 29 August 2019].

3. Espinal, C., & Matulić, D. (2019). Recirculating Aquaculture Technologies. *Aquaponics Food Production Systems*, [online] pp.35–76. Available at: https://link.springer.com/chapter/10.1007/978–3-030–15943–6_3#Sec1 [Accessed 29 August 2019].

4. Goseberg, N., Chambers, M., Heasman, K., Fredriksson, D., Fredheim, A., & Schlurmann, T. (2017). Technological Approaches to Longline- and Cage-Based Aquaculture in Open Ocean Environments. *Aquaculture Perspective of Multi-Use Sites in the Open Ocean*, [online] pp.71–95. Available at: https://link.springer.com/ chapter/10.1007/978–3-319–51159–7_3 [Accessed 29 August 2019].

5. InnovaSea Systems. (n.d.). *Home-InnovaSea Systems*. [online] Available at: https://www.innovasea.com/ [Accessed 29 August 2019].

6. Kumar, G., Engle, C., & Tucker, C. (2018). Factors Driving Aquaculture Technology Adoption. *Journal of the World Aquaculture Society*, [online] 49(3), 447–476. Available at: https://onlinelibrary.wiley.com/ doi/10.1111/jwas.12514 [Accessed 29 August 2019].

7. Luis, A., Campos, E., de Oliveira, J., & Fraceto, L. (2017). Trends in aquaculture sciences: from now to use of nanotechnology for disease control. *Reviews in Aquaculture*, [online] 11(1), 119–132. Available at: https://onlinelibrary.wiley.com/doi/10.1111/raq.12229 [Accessed 29 August 2019].

8. Macchitella, N. (2016). *Smart Floating Farms*. [online] Medium. Available at: https://medium.com/cusp-civic-analytics-urban-intelligence/smar-floating-farms-5cb79866c712 [Accessed 29 August 2019].

9. Marine Insight. (n.d.). *What are Wellboats?* [online] Available at: https://www.marineinsight.com/types-of-ships/what-are-wellboats/ [Accessed 29 August 2019].

10. Martinez-de Dios, J., Serna, C., & Ollero, A. (n.d.). *Computer Vision and Robotics Techniques in Fish Farms*. [ebook] p.1. Available at: https://pdfs.semanticscholar.org/d923/dd001e13c5822cb602c9dec018fc64dcfc6d.pdf%20–1 [Accessed 29 August 2019].

11. Omicsonline.org. (n.d.). *Traditional and Advance Technology in Aquaculture and Fisheries* |. [online] Available at: https://www.omicsonline.org/conferences-list/traditional-and-advance-technology-in-aquaculture-and-fisheries [Accessed 29 August 2019].

12. Phillips, M., Boyd, C., & Edwards, P. (n.d.). *Systems Approach to Aquaculture Management*. [online] Fao.org. Available at: http://www.fao.org/3/ab412e/ab412e13.htm [Accessed 29 August 2019].

13. SEAFDEC/AQD. (n.d.). *Seabass Hatchery-SEAFDEC/AQD*. [online] Available at: https://www.seafdec.org.ph/2011/seabass-hatchery/ [Accessed 29 August 2019].

14. Solon, O. (2011). *Aquapod is a Floating, Sustainable Fish Farm*. [online] Wired.co.uk. Available at: https://www.wired.co.uk/article/aquapod-sustainable-fish-farm [Accessed 29 August 2019].

story-from-bangladesh-2155–9546–1000506.pdf [Accessed 29 August 2019].

19. Van Huong, N., Cuong, T., Nang Thu, T., & Lebailly, P. (2017). Efficiency of Different Integrated Agriculture Aquaculture Systems in the Red River Delta of Vietnam. *Fisheries and Aquaculture Journal*, [online] *8*(4). Available at: https://www.longdom.org/open-access/efficiency-of-different-integrated-agriculture-aquaculture-systems-in-thered-river-delta-of-vietnam-2150–3508–1000230.pdf [Accessed 29 August 2019].

20. Zira, J., Ali, M., Ja'afaru, A., Badejo, B., & Ghumdia, A. (2015). Integrated fish farming and poverty alleviation/hunger eradication in Nigeria. *IOSR Journal of Agriculture and Veterinary Science*, *8*(6), 15–20.

5

Business Planning and Management for Small-Scale Sustainable Aquaculture

CONTENTS

In the chapter business planning and management for small scale sustainable aquaculture, different ideas and approaches that aquaculture firm should use in order to ensure the sustainability in aquaculture is discussed. The chapter explains the role of management in managing an aquaculture enterprise. It also explains the importance of business policies and business system in the management of aquaculture sector. The chapter discusses the five strategies to get the aquaculture growth right. It also explains certain key challenges that aquaculture industry faces in managing the aquaculture activities. In the end, it explains why sustainable aquaculture policies are critical for the growth and development of this industry.

5.1. INTRODUCTION

It is important to note that the aquaculture industry play very important role in maintaining the overall global food security, providing essential nutrition and protein for the population all over the world. One cannot deny the fact that how much important is the aquaculture sector in providing the seafood to all over the world. One of the most fastest growing food sectors, aquaculture has become most vital source in providing fish protein, surpassing the amount of seafood produced for direct human consumption from wild-caught fisheries.

Aquaculture sector is considered environmental favorable when compared it with other food sector, as it is more sustainable than the other animal protein production methods, especially from those production activities that released carbon emissions, the potential for increased farmed seafood production is immense.

Despite the significant growth and development that the agriculture sector is going through, aquaculture is not consequence- or impact-free. With the increase in the number of agriculture firms in the last few years,—often in underregulated or under-managed environments—the industry has experienced both the boom and bust cycles at the same time and attached negative reputation with it because of the environmental impact.

There are certain environmental impacts in the aquaculture sector that are clearly observable such as habitat loss in critical ecosystems (e.g., wetlands and mangroves), nutrient loading that result in poor quality of water, the arrival of invasive species, and the rapid spread of disease and other problems. These impacts may cause severe consequences and can only be addressed by proper planning and efficient management practices in the aquaculture industry. Typically, aquaculture has been developed in an ad-hoc manner, and management has largely focused on licensing, siting, and continuously monitoring of performance in order to reduce the impact of aquaculture activities at the farm level.

This perspective fails to acknowledge that aquaculture industries are dependent on common pool resources (namely land area and water) and are primarily dependent on the ecosystems within which it operates. As there are limited resources available, which result in fish species competing with each other for the resources essential for their survival.

As such, a simple management policy at the farm level will not be helpful in mitigating the risks of negative environmental impacts among all resource users, often proving detrimental to aquaculture industries by creating user conflicts, unable to protect the interest of aquaculture due to the impacts caused by other industries, and diverting the benefits of aquaculture.

5.2. MANAGEMENT OF AN AQUACULTURE ENTER-PRISE

5.2.1. Role of Management

Management in simple terms can be defined as an art to get things done from people. The first and most important function of management is planning, which means setting out a plan to do definite set of things. In other words, it can be defined as setting in advance what to do in future. Before doing a particular task, a manager formulates an idea of how to do that work. It is called planning. The second function of management is organizing. It can be defined as designing of certain set of things performed by each person. In can be defined as managing set of activities, for instance, role performed by each employee and what operations will be carried out by which machinery. Transparency plays very important role in the management to achieve the objective and it is possible only by proper planning.

For instance, in an aquaculture farm, the role of the manager is to clearly specifies the duties performed by technicians, e.g. fertilization, pond liming, and stocking of juveniles. One person is responsible for managing the fisheries stocks, which means how many fish are in the backyard and what is the demand of the product so that to avoid the out-of-stock and over-stock situations.

The third aspect of management is controlling, which means keep an eye on certain set of activities, so that to avoid the duplication of work in order to accomplish the objective of organization. For instance, ensuring standard in the cost of output.

The manager of organization keeps a check on how much total cost incurred by company in producing a certain quantity of goods by considering both capital and operating cost. The total cost is divided by the total quantity and then the cost per unit is calculated. It is then compared with the standard, and in case of there is any difference, then corrective measures are taken by manager.

Controlling is one of the most critical aspects in achieving the objective of organization. There are various activities that controlling manager keep a check on such as schedule, dispatch, prepare, supervise, direct, and undertake corrective measures relative to the specified work.

5.2.2. Business System

In an aquaculture organization, the business system refers to both internal and external factors in which internal factors can be controlled by aquaculture industry while external factors are outside the control of business. Example of internal factors includes conflict between employees and external factors include change of guidelines of government to sell aquaculture products.

The internal variables vary from machineries, raw materials, to money and manpower. In business management, there are 4 Ms. Raw materials are the fertilizers, seeds or fry, and feeds that are vital for the survival and growth of fish. It is the duty of manager to acquire the right material in right quantity and quality, at the right price and the right time for proper scheduling of work in the farm. The machineries used in the aquaculture farm activities include pumps, blowers, laboratory equipment and power generating sets. It is essential that machinery purchased by the production manager must be in accordance with the requirement and specification of plant.

Manpower includes staffs from all the departments weather it is higher authority or lower level employees. In a medium-scale prawn hatchery operation, production staff includes the hatchery aides, head technician, and a phycologist. The production staff is held accountable to the Production Manager who is responsible for scheduling the hours of operations, controlling the quality of input and output of materials, monitors the production activities, and makes sure that there is smooth flow of all task in production department.

Every person is accountable to one another, for instance, Production Manager, in turn, is accountable to the General Manager who is responsible for keeping an eye on entire business operations including marketing and administrative functions. In a similar manner, the general Manager is accountable to the CEO, directors or owners of company.

Every business is established for an aim and that is no other than making profits Without profits, it is difficult for businesses to survive in the long term. The investment put by the owner of company into business is known as owner's equity, while on the other hand, external sources such as debt, loans and trade credits are the sources of financing a business enterprise.

In case of small-scale aquaculture business, it is not easy to finance the business from outside financing through loans from banks and other large financing institution as there are strict rules and regulations and high collateral requirements. The above-mentioned internal variables in the business system are affected by broader forces, e.g., government intervention, technology trends, and market supply and demand.

5.2.3. Business Policy

It is important for every business to have business policy to survive in the long term. A business policy mainly includes various elements and things such as business ethics, objectives, plans, management philosophy, or procedures in the conduct of business. A business policy is composed of a set of principles, rules and action.

An example of a business policy of a hatchery operator is "the company shall endeavor to satisfy the needs of prawn growers by supplying quality fry at reasonable prices." The principle is "satisfy the needs of prawn growers" and the rule of action is "supply quality fry."

A business policy may be concerned with the aspect of making the organization either centralized or decentralized. Centralized organization mean owing of decision-making power to senior personnel, while decentralized organization means distribution of decision-making power to each department or personnel. Most of the aquaculture organization adopt decentralized as it gives power to each individual to exercise his own initiative to a great extent.

A good and effective business policy comprises of the following characteristics:

- It must be designed in such a way that it carries the overall interest of business.

- the goal of the organization must be in accordance with the physical factors, functions, and personnel.

- It must be able to comply with the ethical standards of business.

- It should always be in a clear, simple, precise and understandable terms.

- It should have flexibility and stability.

- It should be sufficiently comprehensive.

- It should be complementary and supplementary.

5.2.4. Business Environment in the Aquaculture Industry

It is important to note that a manager who is responsible for carrying out the aquaculture activities should have good understanding about how the aquaculture industry works, what are the external and internal environmental factors, strength and weakness of company, SWOT analysis, and internal operations of business. it is usually seen that some organization follows a business practice only because it is followed by their competitors. This is not recommended at all. In addition, it is often observed that impressed by the success of neighbor's good business, a company can also decide to go into the same business. This is true in the aquaculture business especially in Panay and Negros Occ. In the southern part of Iloilo, more than 100 small and medium scale prawn hatcheries started after seen the success of other companies. Majority of these investors have started their business after getting influenced by their friends or neighbors who were successful in their initial production runs.

Within the 6 months of starting their business, business owners started facing lot of problems in terms of high operating cost, stiff competition, low survival, high wastage, lack of buyers and lack of cash. All these challenges resulted in closure of business, huge loss, and get into high debt trap. It causes owners to put the blame on their technician or consultant for failure to solve technical problems. They also blame the middlemen for giving low prices for their produce. According to a survey on one startup company, Filipino, it was found that the company failed because of lack of management skill to identify the problems and come out with the best possible solutions. The company also failed to assess the environment conditions in which it was operating.

5.2.5. Aquaculture Industry System

The aquaculture industry generally comprises of system operates in a system composed of groups of interacting and interrelated bodies, with each of them performing an important and vital functions. The central aspect of the system is the production subsystem where mostly all the activities carries on. During the last 15 years, the aquaculture production subsystem has developed into a full- cycle technology for some species, i.e., milkfish, and prawn primarily because of the extensive research undertaken by the SEAFDEC Aquaculture Department.

A full-cycle technology starts from the breeding in captivity of culturable species. In the case of milkfish, the normal duration of the development of brood-stock is about 5 years. With the proper nutrition and right ratio of male to female, it helps in providing the environment of fish mature and spawn. The maturation period of (P. monodon) is 9 to 12 months.

All these activities cannot be performed by the aquaculture firm independently. It requires the support of external forces: government assistance, updated and advanced technology, financing, support industries, physical environment, and market factors. These forces play very important role and success and failure of aquaculture business depend upon these factors.

5.2.6. Corporate Objectives

The corporate objective can be defined as primary purpose for which the aquiculture firm started their business operations. It can be in terms of total profits, overall turnover, and return on investment. For instance, a prawn hatchery operator, estimate total sale for the financial year on the basis present demand and latest trends in the market. In the early 1980's, the market scenario for prawn fry was termed as "seller's market."

This means the seller in the market, known as the hatchery operator, possess most of the power, determine the price of their produce on their own. Later on, as they lack the market information and unaware of the demand of their produce, which resulted in oversupply of fry in the province and the surrounding areas during the 1980's. This further caused fall in the price of their produce, thus affecting the overall sales of the hatchery operators. This low level of demand impacts the overall profit and return on investments of the hatchery operators.

During the formulation of corporate objective, the planning managers should give more importance to these four approaches:

- cost-oriented,
- growth-oriented,
- employee-oriented, and
- community oriented.

The cost-oriented approach is mainly concerned with minimizing the cost of operations in order to increase the cost of business. This is true especially in case of small scale and medium scale aquaculture firms, as they want return on their investment as early as possible.

In the growth-oriented approach, business managers are not interested in short term profits as they want to expand their organization in order to survive in the long term. They set targets with the perspective of long-term achievement. Companies with these objectives keep apart a portion of their profits, so that to reinvest the same over a period of five years. The owners of company with this approach are not interned in return on investment in short term as they are allocating most of their profits for diversification and expansion purpose.

The employee-oriented corporate approach view employees as the asset of organization. this approach gives employees higher importance in the company and give them enough opportunity for their growth and development. This approach also provides them job security, favorable working conditions, and fair management-labor relationship in order to achieve the objective of organizations.

The community-oriented corporate objective gives more importance to the relationship of the organization with the immediate community. The company with this objective always pays fair taxes to the government, creates business opportunities for others, provides employment, helps in preserving the ecology, works for the social cause and supports community activities. In short, the company helps in the economic uplift of the community.

5.3. SUSTAINABLE FISH FARMING: FIVE STRATE-GIES TO GET AQUACULTURE GROWTH RIGHT

Figure 5.1: providing good quality feed to fish.

Source: https://encrypted-tbn0.gstatic.com/images?q=tbn:ANd9GcRUETR1B NwJzrDlq4ljWPzMW8efebeCm_knwOgJQoNuUAByW1Im

Within the past few years, due to the increase in the population and change in the consumption pattern of people, there has been significant increase in the consumption of fish. As people are becoming more active about living a healthy life, they started consuming fish as it is a vital source of protein. Shellfish and finfish currently make up one-sixth of the animal protein people consume globally.

As the global wild fish catch peaked in the 1990s, aquaculture or fish farming has been increased significantly in order to fulfill the demand of world population. According to recent research, it was observed that there will be needed to increase aquaculture production two times between now and 2050 to fulfill the demand of increased population.

One of the important questions in the sustainable fish farming is: is it possible to grow aquaculture sustainably?

5.3.1. Aquaculture's Impacts: Encouraging Trends, But Challenges Remain

On average, farmed fish are very important for the overall world population as it provide vital source of protein, thus it provides enough feed as efficiently as poultry, making them an attractive option for expanding the global animal protein supply. However, just as it is the case with all forms of food production, aquaculture activities cannot be possible without its environmental impacts.

As aquaculture began to boom in the 1990s, there are various issues raised along with this such as the clearing of mangroves for the development of shrimp farms in Latin America and Asia, increased use of fishmeal and fish oil primarily available from wild marine fish, resulted in increase in the shrimp and fish diseases and water pollution.

There has been significant growth observed in the aquaculture industry within the past 20 years, producing more farmed fish per unit of land and water, which resulted in reducing the share of fishmeal and fish oil in many aquaculture feeds, and largely stopping the number of mangrove conversion.

However, doubling the aquaculture production, without increasing the efficiency of aquaculture industry could result in doubling of environmental impacts. And unless there will be boost in the productivity of aquaculture industry, the limited availability of land, feed and water may constrain its growth.

5.3.2. Getting Aquaculture Growth Right: Five Approaches

There are five approaches that help in getting the aquaculture approach in the right direction:

Invest in technological innovation and transfer

Aquaculture is a new industry and started gaining importance with the increase in demand of fisheries products. As there is so much expectation from this industry, there is a need to improve the technology in order to fulfill the demand of current population as well as getting prepared for the future generation. New innovation in the aquaculture sector helps ensuring disease control, feeds and nutrition, breeding technology, and development of low-impact production systems.

These sorts of innovations—whether led by research institutions, farmers, companies, scientists, or governments—have been behind productivity gains in every part of the world. For example, in Vietnam, a breakthrough in catfish breeding around the year 2000—accompanied by the high-quality feed to the fisheries—opens the door for significant growth and intensification.

Vietnamese catfish production grew from 50,000 tons in the year 2000 to more than 1 million tons in 2010, even though the total area of catfish pond in the country only doubled during that time.

Focus beyond the farm

It is usually seen that most of the rules and regulation designed for the aquaculture sector are valid at individual farm level. It means there are no standard guidelines that are applicable at national or world level. Thus, by not having strict guidelines and rules, it may cause significant impact on environment as producers will be negligent in performing their duty towards environment.

Most aquaculture regulations and certification schemes focus at the individual farm level. This may cause water pollution of fish diseases, which impact the entire food chain. Zoning and spatial planning can ensure that aquaculture operations stay within the surrounding ecosystem's carrying capacity and can also lessen conflicts over resource use. For instance, Norway's zoning laws, are responsible to keep a check that salmon producers are not overly concentrated in one area, which helps in mitigating the environmental impacts and reducing disease risk.

Shift incentives to reward sustainability

Within the past few years, concerning the increase in the environmental problems, there are many public as well as private policies designed to give incentive to farmers engage in practices of sustainable aquaculture. For example, the government of Thailand give opportunity to shrimp farmers to operate legally in aquaculture zones with access to water supply, free training, and wastewater treatment.

The government has also provided tax exemptions and low-interest loans to small-scale farmers—allowing them to use advance technology that enhance their productivity, reducing pressure to clear new land.

Leverage the latest information technology

Advances in satellite and mapping technology, open data, ecological modeling, and connectivity mean that planning systems and global-level monitoring that encourage the practices of sustainable aquaculture may now be possible.

A platform designed with the help of these technologies allow the governments to improve monitoring and spatial planning, help the industry plan for and demonstrate sustainability, and help civil society to be successful in their field and hold government and industry accountable for any wrongdoing.

Eat fish that are low on the food chain

According to various studies and researches, it was found that in the fish farming, the overall pressure on the marine ecosystems can be reduced, if farmed herbivorous or omnivorous fish—"low-trophic" species such as catfish, tilapia, bivalve mollusks, and carp.

In emerging economies, where there is still high demand of low-trophic species, emphasis should continue with these species even as billions of people enter the global middle class in coming decades. At the same time, as fish provide necessary source of proteins for billions of people in developing and less developed countries, growing aquaculture to meet the food and nutritional needs of these consumers will be essential.

With the global wild fish catch stagnant and the world population is continuously increasing, aquaculture is here to stay. Thus, there is a need to grow the aquaculture sector in right direction and ensure that fish farming contributes to a sustainable food future.

5.4. KEY CHALLENGES IN MANAGEMENT OF AQUACULTURE

Figure 5.2: Challenges in maintaining the quality of water in aquaculture sector.

Source: http://www.arc.agric.za/arc-api/SiteAssets/Pages/Aquaculture/Picture1.jpg

Within the past few years, with the rise in the aquaculture activity, there has been significant increase in the number of challenges that aquaculture sector is facing. These problems are closely associated with the geographies and production system. Based on various studies conducted all over Indonesia by BAPPENAS, SFP, CI, and partner universities, some of the key issues that the management of the aquaculture firms face are discussed in the following subsections.

5.4.1. Conflicts with Other Resource Users

It is usually seen that aquaculture firms requires the allocation of public space (land, marine area, coastal, or freshwater) and may result significant habitat modification or conversion. It may cause significant direct and indirect impacts on all resource as the functioning of one is dependent on others.

It creates disruption in use of resources in case the rights and equity are not properly accounted for. Most of the conflicts that arise with the resource users are only because of inadequate protection of high-value and sensitive ecosystems.

6

Rural Aquaculture Impact on Livelihood and Food Security

CONTENTS

There are around two billion people around the world that are directly or indirectly dependent on fisheries for livelihood and food security. The problems of malnutrition and lack of essential nutrients is rampant in developing countries and solution is provided by fisheries.
The fishes and aquatic animals are biggest source of animal protein and essential nutrients that can fulfill the essential requirements of human bodies. The aquaculture also earns the developing countries much needed foreign exchange and very good price to fishers for their catch.

6.1. INTRODUCTION

The intensive culture of salmon in developed countries and rearing of shrimp in developing countries; that are done by particularly by good farmers to supply a high-value product for rich consumers, generally with negative environmental impact are all come under the definition of aquaculture.

The poverty ridden farmers farm fish in regions of Asia where it is old-style practice; they eat or sell fish to earn income; have confidence in the technology; and have reach to appropriate genetic material that is fish seed. The utilization of sharing instead of technology-driven methods is causing the adoption of aquaculture by new entrant poor farming houses in Africa and Asia.

The farming of fish and other aquatic organisms is defined as aquaculture. All farmed aquatic organisms are in general terms called as fish. The rearing of fish in rice fields and ponds is generally combined in agriculture in land base systems. The keeping of fish directly in enclosures or attaching them to substrates in water bodies such as rivers, lakes, reservoirs or bays is contained in water based systems.

The starting point for landless people and poor fishers to farm fish is involved in Water-based systems. The low-cost production with extensive and semi-intensive technologies most suitable for the restricted resource base of small level families is defined as Rural aquaculture according to Edwards and Demaine in the year of 1997.

It depends on natural food for fish, for example, tiny fish, which is every so often prepared and additionally enhanced by other feed. Manure and feed might be gotten from on-farm side-effects, at any rate in the underlying phase of intensification, albeit even here created and pelleted feed from agro-industry are progressively utilized. On the other hand, intensive systems perpetually rely upon moderately high expense, healthy complete food diets. The yearly development of aquaculture has grown at average of some 11% for almost two decades.

Though salmon and shrimp get most attention, they include under 10% of worldwide aquaculture production by weight in comparison with 50% for carps and tilapias that contribute most to the local food supply in developing countries nations. The Aquaculture production is additionally slanted topographically with Asia delivering over 90% of products worldwide, outweighing Africa and Latin America at under 0.5% and 2% respectively.

The 67% of aquaculture production around the world is done by china, inland aquaculture production has expanded at minimum fivefold in the previous decade; it has just doubled in the rest regions of the world, inferring huge potential in other developing countries if restrictions to its extension were removed.

From the decade of 1970 the aquaculture production has thrived quickly, and was the quickest developing food manufacturing industry in numerous nations for as far as two decades, surpassing terrestrial farm animal meat production and arrivals from catch fisheries (Tacon, 2001). From the many past centuries, aquaculture has been started in many regions of developing countries, for example, Africa and Asia, with the goal to open up open doors for the nearby provincial communities to improve their way of life and an approach to escape from poverty (Edwards, 2000).

It is accepted that aquaculture is probably the quickest ways for the poor to procure a living while act as a valuable foreign exchange for the national advancement. In addition, aquaculture has gradually incorporated as a fundamental piece of village livelihood when it turned out as the answer for strengthening population pressure environmental degradation or loss of access, the decrease in catches from the wild fisheries (Halwart et al., 2003).

While the worldwide wild catch decreased at the rate of more than 0.5 million tons for each year, aquaculture has been developing at generally 2.5 million tons every year between 2004 and 2011 as per Cleasby in the 2014. The village aquaculture is generally explained as aqua farming practices in extensive to semi-intensive scale with generally low production cost and technology (Edwards & Demaine, 1997).

The targeting of low-salary shopper groups, this small-scale household activity adopted off-farm Agri-industrial inputs and organic fertilizer without depending on any formulated feed to supply less value production. Aquaculture improvement is frequently part of the rural advancement program since the majority of the aquaculture was broadly promoted in rural regions.

Though the advancement referenced here is uncertain and exceptionally disputable, which whether it pursues its conventional polarity between country or agrarian and urban or industrial regions (Yap, 1999). It is by no means a simple assignment to set up an effective aquaculture industry, also in a village area where access to assets is truly restricted.

There are several real cases till today, occurred in for the most part Africa and some other underdeveloped countries, where introduced aquaculture has not been successful. The general example of this can be the cage culture of carp in Bangladesh. The cage culture was formally acquainted with Bangladesh in the decade of 1980 in the Kaptai Lake (Ahmed and Saha, 1996; Edwards et al., 2002). The carp industry soon collapsed due to the inability of local people to inject capital cost and inputs to afford cages according to Bulcock et al. (2000). The absence of access to capital and resources, vulnerability, and aversion in villagers to take the risk is the reason behind an aquaculture failure (Asian Development Bank, 2005).

The lack of specialized skills and learning in operating fish farms by the local village people are also an important obstacle to overcome. The inadequate monetary help from the administration and the residents has put village aquaculture to a more troublesome spot (Edwards et al, 2002).

Though in majority of the time aquaculture farms operated by the village families are in the mission of improving the life standards in context of removing poverty and securing food availability, the successful aquaculture cannot make sure the previous terms and also not comes along with all the facilities. The village aquaculture can also be a weapon that threatens the poor without careful management. Every coin has two faces. The aquaculture done by the village communities does bring bad results as well in addition the benefits to the villagers.

6.2. IMPACT OF AQUACULTURE ON THE LIVELI-HOOD

The poor people livelihood is supported by agriculture through improved food supply, employment and income. There are many small level farmers that have small land possessions in regions of complex, diverse and risk prone agriculture in mainly rainfed and undulating arrive on the edges of marshes or in uplands. The development of a pond a pond on these regularly environmentally degraded farms may also give a focal point to agriculture diversification and increased sustainability, by giving a source of water.

The poor in well-endowed lowlands are generally landless or close landless; the fish cultivating in common water bodies may help to decrease poverty, provided that the poor can get to them. The Inland and coastal front fishers are generally landless and are among the most devastated; their chances lie essentially with water-based culture systems. Despite the fact that fish give far less animal protein to world nutrition than livestock, the people in major areas of Africa and Asia are highly reliant on fish as a major aspect of their day by day diet in 18 countries in Africa and Asia, nine on every continents, fish provide at least 40% of dietary animal protein. It also gives highly digestible energy, and is a good source of fat and water solvent nutrients, minerals and unsaturated fats. Aquaculture has contributed in the past towards decrease in poverty in poor society in the couple of regions of the world in which it is conventional practice, for example China, Indonesia and Vietnam, and it keeps on doing as such today. There are very less products that have explicitly focused on poor people and the effect of aquaculture on the poverty has hardly been evaluated.

The Recent experience in Asia and Africa indicates that poor farmers adopt aquaculture where certain predisposing conditions are met:

- The farmers and Consumers both must see the price of fish and this must be reflected in market demand. This is most possibly be the situation in regions where wild or cultivated fish have generally been devoured. Early endeavors to encourage aquaculture regularly focused around house hold food security that might have added to the high rate of failed improvement activities, especially in Africa.

The methods focused on have the more chances to be more appropriate

- For the poor to become legitimately related with aquaculture as a farming practice, they have to own or lease horticultural land for culture in rice fields or lakes, or if landless have access to water body in which they can stock fish.

- The another main need is learning of proper technology, for example some small scale farmers failed to culture fish in northeast Thailand as they supplied too few fingerlings at too high a thickness in lakes that were neither fertilized nor fed; there are high death rates of fish; were consumed by left populations of flesh eating wild fish, or simply did not grow according to Edwards in the year of 1996.

- A supply of seed is vital and is regularly the main restricting element for selection of aquaculture. The government support is generally required for new entrant farmers in the form of extension advice, inputs, especially seed.

6.2.1. Potential Contribution of Aquaculture to The Livelihoods of The Rural Poor

Direct Advantages

- The high nutritional value of Food, particularly for vulnerable groups, for example, pregnant and lactating ladies, babies and pre-younger students.

- The own business employment including for ladies and youngsters.

- Income through sale of produce or yield that has a moderately high value

Indirect Advantages

- The more availability of fish in local market area and in urban markets, that might cut costs own.

- The jobs on bigger farms, in seed supply systems, market chains and production or manufacture capacities.

- Benefit from normal pool assets, especially the landless, through cage culture, culture of mollusks and seaweeds, and upgraded fisheries in collective water bodies.

- The greater farm sustainability through development of ponds which additionally serve as little scale, on-farm reservoirs rice and fish culture as a component of integrated pest management.

6.3. IMPACT OF AQUACULTURE ON THE FOOD SECURITY

According to Food and Agriculture Organization the Food security is defined as the situation where all the people living, at all times, have physical, social and economic reach to sufficient, safe and nutritious food that fulfill their dietary needs and food preferences for an active and healthy life as explained by Schmidhuber & Tubiello (2007).

The Aquaculture in small farmer system in village areas gives a high quality of animal protein and primary supplements, particularly for nutrition vulnerable groups, for example, pregnant and lactating ladies, babies, and pre-younger students. There are nearly 50% of the child deaths around the world are connected to malnutrition. In numerical context, it is around 3 million young lives each and every year according to UNICEF statistics (2014–2015). It was demonstrated that after provided with adequate required nourishment which can be found in fishes, for example, vitamin B12, calcium, and potassium, disastrous cases like child visual blindness and newborn child mortality has substantively reduced according to Ahmed and Garnett (2011).

In the village areas, more often farmers family will in general eat the little fish that does not meet the market size and left the greater one which can bring more expensive rates (Ahmed and Garnett, 2011). Once in a while, there are some village communities do practice by giving out fishes as a sort of payments to workers working in the houses (Irz, 2007). These little fishes are eaten together with their head and bones, it contains more micronutrients, vitamins and mineral that could not be found in bigger fish (Ahmed and Garnett, 2011).

Though in the other way the act of collecting free fish from ponds has contributed as the basic nutrient source to the poor families in rural areas, and aided in decreasing malnutrition among young kids. The contribution of aquaculture to food security to the general wellbeing betterment were obviously shown in Pacific Island Countries and Territories through diversification to tuna farming. Aside from the financial benefits, the tuna farming is great in such a manner that assisted in fighting the high and rising prevalence of non-communicable disease of the local individuals over the area (Bell, 2015).

The non-communicable diseases like heart disease and obesity van also happen to poor people. The greater dependency on imported and processed

2001). According to FAO (2001), the significant, though frequently ignored, advantage which is especially applicable for incorporated agribusiness aquaculture frameworks, is their commitment to expanded farm effectiveness and manageability.

Farming side-effects, for example, compost from domesticated animals and harvest buildups, can fill in as manure and feed contributions for little scale and business aquaculture. Fish cultivating in rice fields adds to coordinated bug the board, yet additionally the board of vectors of human restorative significance (Halwart, 2001). Moreover, lakes become significant as on-ranch water stores for water system and domesticated animals in zones where there are occasional water deficiencies (Lovshin, 2000).

In perspective on every one of these advantages, it is maybe not astounding that aquaculture creation has developed quickly from the decades of 1970, and has been the quickest developing food production area in numerous nations for about two decades; the division displaying a general development rate of over 11.0% every year since 1984, contrasted and 3.1% for territorial farming fish meat production, and 0.8% for arrivals from catch fisheries (Tacon, 2001).

By 1999, the production of all refined amphibian life forms achieved 42.8 million metric ton (FAO, 2001). An aggregate of 262 fish, scavenger, and mollusk species, speak to the most significant creatures utilized in aquaculture around the world, are recorded in an ongoing study as per Garibaldi in 1996. In spite of the fact that not every amphibian living being are appropriate for culture, the assortment of refined species is as yet expanding. Freshwater finfish, especially Chinese and Indian carp species, represent the best portion of complete aquaculture creation in 1999. This is trailed by mollusks and oceanic plants, for the most part kelp, most of which originate from China. FAO's most recent investigations on future interest for, and supply of, fish and fishery items anticipate a sizeable increment sought after for fish (FAO, 2000).

The majority of this growth will result from expected economic development, population growth, and changes in eating habits. Fish supply from capture fisheries in most countries is expected to remain constant, or decline, since catches have either reached, or are close to, maximum sustainable yield. Inland fisheries may yet be able to yield more fish as effort increases, but the increased effort required will become increasingly challenging.

The inside the country's fisheries are also in danger to environmental effects, such as watershed degradation, development of water control structures and pollution. The characteristics of the changing village environment. The aquaculture has an important role to play in meeting the growing requirement demand for fish. Certainly, the increase of global aquaculture is forecasted to continue for some time (FAO, 2000).

REFERENCES

1.   Allison, H. E., & Ratner, B. (n.d.). *Aquaculture, Fisheries, Poverty and Food Security*. [ebook] malayasia: The WorldFish Center, p.2. Available at: http://pubs.iclarm.net/resource_centre/WF_2971.pdf [Accessed 29 August 2019].

2.   Edwards, P. (2000). *Aquaculture, Poverty Impacts and Livelihoods*. [ebook] Natural Resource perspectives, p.2. Available at: https://www.odi.org/sites/odi.org.uk/files/odi-assets/publications-opinion-files/2849.pdf [Accessed 29 August 2019].

3.   Halwart, M., Funge-Smith, S., & Moehl, J. (n.d.). *The Role of Aquaculture in Rural Development*. FAO, p.2.

4.   Kar Mun, A. (2017). *Impact of Aquaculture on the Livelihoods and Food Security of Rural Communities*. [online] Available at: https://www.academia.edu/35280580/Impact_of_aquaculture_on_the_livelihoods_and_food_security_of_rural_communities [Accessed 29 August 2019].

5.   Olaganathan, R. (2017). *Impact of Aquaculture on the Livelihoods and Food Security Of Rural Communities*. [online] Research Gate. Available at: https://www.researchgate.net/publication/316123854_Impact_of_aquaculture_on_the_livelihoods_and_food_security_of_rural_communities [Accessed 29 August 2019].

6.   Rajee, O., & Kar Mun, A. (2017). *Impact of aquaculture on the livelihoods and food security of rural communities*. International Journal of Fisheries and Aquatic Studies, p.2.

7

Application of Biotechnology in Rural Aquaculture

CONTENTS

Biotechnology is an important field which has several applications in aquaculture. The importance of biotechnology and role of biotechnology in improving rural aquaculture has been explained. The need for the application of biotechnology in aquaculture has been described. There are several techniques of biotechnology which are used in rural aquaculture and these techniques have been elucidated. There are several challenges in the utilization of biotechnology principles in aquaculture and these challenges have been mentioned. The latest techniques and emerging techniques in biotechnology that are used for the betterment of aquaculture have been described.

7.1. INTRODUCTION

Biotechnology refers to any technology that uses the principles of biology and it is one of the most advanced branches of science that is developing rapidly in the recent times. Biotechnology has important principles and techniques that are used for the development of different areas such as aquaculture, fisheries and in certain fields related to food industry.

There is a significant increase in the demand for seafood and fishes with the increase in the population and there is a decline in the natural marine habitats because of this increasing demand. Therefore, there are several methods and techniques that are being used to increase the production of aquatic organisms to cater the needs of the ever-growing population.

Biotechnology is the field which facilitates scientists to combine different genetic traits of various organisms by using the principles of molecular biology and genetic engineering, to improve the productivity and quality of the food that is being produced by aquaculture methods. There are several scientists belonging to biotechnology who are working consistently to implement advanced techniques of science and technology to aquaculture and improve the quality and productivity of the food that is obtained from aquaculture.

There are some researchers who are trying hard to find some natural fish growth factors that can improve the disease resistance in different fish. The novel techniques and innovations in biotechnology are contributing to the development of aquaculture and fisheries. The utilization and implementation of these novel methods is facilitating the improvement of aquaculture by resolving the hindrances that are limiting the growth of aquaculture.

There are different challenges, concerns and limitations that are preventing the growth of aquaculture and these cannot be resolved by the use of traditional methods. Therefore, there is need for novel and advanced technologies to solve the problems that are related to the growth and development of aquaculture which are being solved by the biotechnological methods.

The utilization of novel methods of biotechnology is known to improve the productivity of the aquatic species and help in meeting the demands of the growing population by improving the productivity of aquatic species through aquaculture practices. The techniques and principles of genetic engineering and biotechnology have an enormous scope and potential to improve the quality and yield of fish that are produced by aquaculture.

The demand and need for agriculture are increasing and these demands for sea food through aquaculture can be met only by the principles of biotechnology. However, there is need for proper policy making and regulation for release of foods that are obtained by biotechnology before these foods are released into the market.

The aquaculture practices that involve biotechnology are known to solve several problems related to environment. There are some traditional aquaculture practices which are known to have a number of adverse hazards on the environment and these problems can be resolved by the application of biotechnology to aquaculture. If the techniques of biotechnology are integrated with aquaculture and are well implemented then the demands of increasing population can be easily met and these practices enable to increase the food availability and food security in different nations across the globe.

The successful implementation of techniques and principles of biotechnology is generally possible only by having vast knowledge in the fields associated with biotechnology such as biology, breeding, physiology, biochemistry, genetics, pathology, agronomy and variation in different organisms. There can be improvements in the field of biotechnology by continuous research and good command over the fundamentals of several fields associated with biotechnology. Therefore, it is important to understand that continuous improvement of knowledge and relentless research in biotechnology is the key to the success of the branches associated with biotechnology.

Aquaculture and fisheries are known to be an important area that is associated with the production of food, assuring nutritional security, increasing food availability and establishing food security. The aquaculture a sector that is known to have importance across the world and it is known

biological systems or living organisms in production process according to has a wide range of useful applications in fisheries and aquaculture.

The research and development in biotechnology are growing at a rapid pace. Biotechnology has a significant role in improving the techniques and procedures used in the development of fisheries, human health and agriculture. The use of biotechnology for the improvement of aquaculture led to the development of new tools and techniques that have the capacity to create several novel genes and genotypes of plants, animals and aquatic species.

The use of biotechnology for the improvement in aquaculture and fisheries is known to be a novel field but it has significantly improved the production of fish and other organisms by aquaculture through the application of biotechnology.

Biotechnology is considered as an important area which is associated with the growth and development of fisheries and aquatic organisms. Biotechnology is known to be instrumental in improving the production of fish and implementation of the techniques of biotechnology is known to revolutionize the areas of aquaculture and fish farming.

Biotechnology is known to have an important role in the conservation of biodiversity. The important techniques such as the utilization of synthetic hormones in fish breeding, transgenesis, chromosome engineering and application of biotechnology in improving the health of the aquatic organisms and better gene banking methods has been identified.

7.3. TECHNIQUES OF BIOTECHNOLOGY USED IN RURAL AQUACULTURE

The important branches of biotechnology that are widely used in aquaculture are known to be utilization of several synthetic hormones that are used in induced breeding, production of monosex, uniparental, and polyploid population. There techniques of genetic engineering and molecular biology that are sued in the synthesis of transgenic fish, improvement in gene banking, better health management and development of several natural products from biotechnology.

7.3.1. Biotechnology in Fish Breeding

The fish breeding in several fish can be improved by the techniques of biotechnology and Gonadotropin Releasing Hormone (GnRH) is known to one of the most important hormones that is used along with principles of biotechnology for improvement of fish breeding in several place. The Gonadotropin Releasing Hormone (GnRH) is known to be an important regulator and inducer of reproduction in several vertebrates (Bhattacharya et al., 2002).

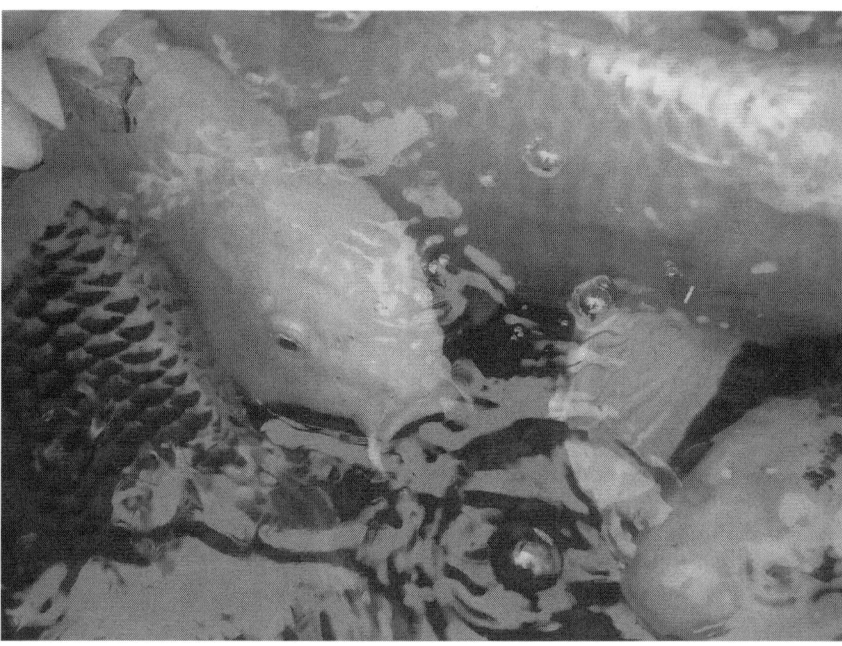

Figure 7.2: Representation of breeding fish.

Source: https://www.maxpixel.net/static/photo/1x/Breeding-Fish-Red-Koi-Colored-Carp-Aquarium-Fish-

1447284.jpg

Gonadotropin Releasing Hormone (GnRH) is a hormone which was initially isolated from the hypothalmi of the pig and sheep. This hormone is a decapeptide and it is known to be responsible for releasing the luteinizing hormone (LH) and follicle stimulating hormone (FSH) which is two important hormones that play an important role in reproduction (FSH) (Schally et al., 1973).

GnRH is known to be the only hormone which is widely produced in the placenta of mammals and human beings that is known to be involved in triggering reproduction. GnRH is a neuropeptide which is known to release the LH and FSH hormones which have a role in reproduction. There is only one form of this hormone that is being isolated from all the mammals from the time of its discovery.

There are twelve variants of the GnRH that is present in non-mammalian species except guinea pig. These twelve variants of GnRH have been elucidated structurally and among all these twelve variants there are nearly seven or eight forms that have been isolated only from fish (Halder et al., 1991; Sherwood et al., 1993; King and Miller, 1995; Jimenez Linan et al., 1997).

The most recent variant of GnRh that has been characterized and purified were known to be isolated and characterized by the group Carolsfeld et al. (2000) and Robinson et al. (2000). There are several structural variants that have been isolated along with their biological activities and different types of chemical analogues have been synthesized and are widely used in the breeding of fish. These commercial variants of GnRH that are sued in the breeding of fish have a commercial name called as "Ovaprim" that is used across the world.

There are some commercially important fish which are present mostly in land locked water cannot reproduce until they are induced by this Gonadotropin Releasing Hormone (GnRH).The technique of induced breeding is known to be a recent development in the field of Gonadotropin Releasing Hormone (GnRH) technology and it is being widely used for the improvement of the production of food.

7.3.2. Fish Feed Biotechnology

The most commonly used source of food for many fish is known to be the fish meal and fish meal is known to be a product that is produced by processing fish to produce a high-quality high protein content that can be used as food for several fish.

There are many advantages by providing fish feed as food for several fish but along with the advantages there are several disadvantages. One of the major disadvantages related to the production of food is that the fish produce is extremely expensive which would make the fish produced by the aquaculture also to be extremely expensive. The other disadvantage is that the fish meal that is produced from fish is known to be obtained by processing of several verities of wild fish and there are studies that indicate that there is huge decline in the availability of fish across the world. In addition to these there is another disadvantage with respect to the use of fish fee which is that the use of fish feed is known to create a lot of environmental pollution and other problems associated to the environment.

There are some fish feed which might contain small sources of phosphorous which is slightly above the requirement of the fish and there are chances for small quantities of this phosphorous top enter into the water. This entry of unnecessary chemicals into the water may lead to eutrophication that is responsible for excessive growth of algae and might also lead to the biomagnification which has several negative effects on the environment.

The presence of these concerns with relation to the production of fish fee there are several methods of biotechnology that are being used for the production of plant-based protein source. The plant based protein that is being synthesized with the help of biotechnology is known to resolve several problems that are associated with phosphorous pollution and is known to be ecofriendly by being beneficial to several organisms.

7.3.3. Bio-Remediation

There are several species of fish and other aquatic species that are easily affected by disease when they are cultured. The occurrence of disease in several species of aquatic organisms is known to have huge environmental impact.

The major reason for occurrence of disease in several aquatic animals is that these culture species are known to have lesser disease resistance when compared to the wild species and they are susceptible to different diseases. The farm species of fish and other aquatic species are known to a lot more susceptible to the changes in the environment along with their disease susceptibility. These aquatic species are grown in fresh water ponds and other pond on which they are extremely dependent on for the oxygen, important chemicals, uptake of waste products and they can easily carry pollution from the nearby environment. (Mandany et al., 1996).

The development in biotechnology and several techniques in biotechnology are widely used in development of aquaculture for bioremediation as aquatic species have a close relationship with the surrounding environment. This close relationship that is present between the aquatic species and environment led to the development of a novel field in biotechnology which is called as the bioremediation. The process of bioremediation refers to the use of healthy good bacteria to treat the water or fish feed by the use of natural processes to improve the ponds by removing the harmful pollutants and disease-causing bacteria that are present in the aquaculture farms (Verschuere et al., 2000).

7.3.4. Biotechnology and Fish Health Management

There are a number of species of fish which face a major problem with respect to disease resistance. A number of aquatic species that are being cultured are known to be extremely susceptible to diseases. There are several methods that are being used to improve the disease resistance in aquatic species and the implementation of certain biotechnological tools is known to play an important role in improving the disease resistance of several aquatic species that are being cultured.

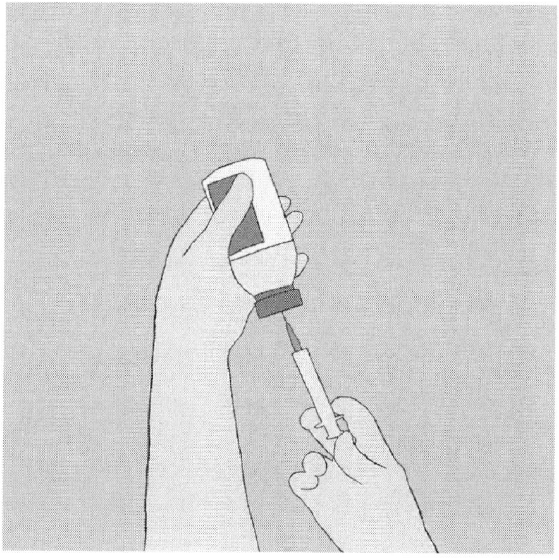

Figure 7.3: Vaccines or immunostimulants used for disease resistance.

Source: https://cdn.pixabay.com/photo/2017/09/26/15/08/vaccina-tion-2788900_960_720.jpg

There are several techniques such as molecular diagnostic methods, vaccines and other immunostimulants that are being implemented in aquaculture to improve the disease resistance in fish and shell fish across the world.

There are some important biotechnology methods such as use of gene probes and PCR which are popularly used in diagnosis. These diagnostic methods have been designed and developed against some pathogens which affect the growth of fish and shrimp (Karunasagar, 1999).

The biotechnology tolls and techniques facilitated in development of several vaccines against bacteria and virus in several finfish in aquaculture. Some of these techniques of biotechnology used in disease resistance are known to contain few traditional vaccines that can kill microorganism. However, novel vaccination methods like subunit vaccines that contain genetically engineered organisms and DNA vaccines have been developed with the help of biotechnology.

The process of vaccination and immunization for disease control is an important feature that is common to several vertebrates. The shrimps are known to be those organisms which have an extremely weak immune system and the techniques of biotechnology are used for the development of a single molecule that can stimulate the immune system of shrimp.

There are several studies and research that is being carried in the area of disease resistance which have indicated that there is a nonspecific defense system that can be stimulated by using microbial products such as lipopolysacharides, glucans or peptidoglycans (Itami et al 1998). There are several immunostimulants that are present among which glucan and levamisole are known to increase the specific antibody responses and improve the phagocyte activities (Sakai, 1999).

7.3.5. Transgenesis

The process of transgenesis is a technology associated with the biotechnology and this field is widely concerned with the improving the genetic traits of several commercially important fishes, mollusks and certain crustaceans that are known to be important for aquaculture.

Figure 7.4: Representation of transgenic fish.

Source: https://live.staticflickr.com/3003/2454306882_0ab1af64c2_b.jpg

The process of introduction of some exogenous gene/DNA into the host genome resulting in stable transmission, maintenance and expression of the genes is called as transgenesis. There was a technique related to transgenics that was developed by Danish et al and this technique is known to be successfully applied for several species of fish.

This technique of transgenesis developed by Danish et al is known to have resulted in significant improvement and growth of certain aquatic species such as Salmonid. This technique offers a lot of scope and potential in improving and modifying several genetic traits of some important aquatic species such as fishes, crustaceans and mollusks. This is an important method that is used in aquaculture for the improving of production of fish and other aquatic species.

7.3.6. Cryopreservation of Gametes or Gene Banking

Cryopreservation refers to a technique which involves the long-term storage of several biological materials at an extremely low temperature. This technique refers to the storage of several biological organism or material with the help of liquid nitrogen at nearly –196°C.

The principle of cryopreservation is that extremely low temperatures would immobilize or tranquilize the biochemical and physiological activities of the cell and make the cell to survive at low temperatures for extremely long periods.

Figure 7.5: Image of liquid nitrogen which is used in cryopreservation.

Source: https://live.staticflickr.com/5251/5544588680_4309fa6010_b.jpg

There are several animal husbandries that have adopted the cryopreservation of the fish spermatozoa. This technique is being used for animal husbandries for extremely long periods. The initial success related to the preservation of the fish sperm to low temperatures was recorded by Blaxter (1953). Blaxter found success in cryopreservation when the eggs of *Clupea herengus* (Herring) were fertilized with the help of thawed semen.

The cryopreservation technology in recent times facilitated the preservation of spermatozoa of several cultivable fish (Lakra, 1993). There are several advantages related to the cryopreservation technology as this technology would enable in overcoming the hurdles related to maturation of male before the female. This method facilitates selective breeding and leads to the stock improvement related to conservation of species.

One of the major advantages of the cryopreservation technology is that it enables breeders to develop new strains by fertilizing different organisms. There are some limitations with respect to the cryopreservation of the aquatic gene bank as the male gametes of finfishes can be stored and there is no viable technique that can freeze or cryopreserve the eggs or embryos of finfishes. There is a recent development in the area of preservation of embryos where the shrimp embryos could be carefully preserved. The embryo preservation techniques are known to be important advancements in the field of cryopreservation and genetic engineering.

7.3.7. Microbial Technology for Aquaculture

The aquatic environments are known to contain several microbial communities which are part of their environments and these microorganisms are known to have an important role in food webs and organic mineralization. These microorganisms belong to different categories and they are known to have a role in the nutrient and energy flow patterns.

There are several aspects related to microbial technology in aquaculture as it is associated with several aspect of aquaculture such as bio fertilization, microbial processing of organic matter, improved technologies in feed digestibility, detritus enrichment and other issues related to the benefit of human beings were known to use techniques of microbial technology.

Microbial technology also used for reducing the length of the food chains and improves the energy transfer rates by using techniques of biotechnology along with principles of microbiology. There are several techniques that are used for the up gradation of bacterial strains, bio filtration and waste recycling which were used for enhancing the productivity and quality of the aquatic organisms (Ayyappan, 1994).

7.4. CHALLENGES

There are several techniques related to biotechnology such as the production of transgenic fish which are generally made by generating a lot of change in the organisms. There are a number of concerns that are present in relation to biotechnology of aquatic concerns and some of the main concerns are associated with the environmental damage that would be caused by these organisms in the future.

There is another important concern regarding the transgenic fish which refers to the public perception of the transgenic fish and other foods. The public perception of the transgenic fish and consumption of genetically modified organisms is a major concern. There may be several misconceptions and doubts regarding the consumption of the genetically modified fish among the common people because of which there would several people who would not be interested in consumption of the transgenic fish. Therefore, there is need to create awareness regarding the novel technologies and their acceptance among people to obtain easy acceptance.

A number of techniques of biotechnology need a lot of investment and sophisticated laboratories to implement them which becomes extremely difficult for some nations to support these research programs as they are expensive.

There may be concerns regarding the transgenic fish that is created by the biotechnology as the impacts of this transgenic fish when they breed with the wild species is not known and the magnitude of damage this can caused cannot be estimated because of which several transgenic animals are made sterile. This sterility that is induced in fish is mostly reversible and it is possible by the injecting a synthetic hormone that can induce breeding in these fish.

One of the major challenges related to the biotechnology in aquaculture is that there is need for careful planning and budget that needs to be preset for conducting field trials because of which there is need for careful cost evaluation. If there is no prior planning and estimation then the conducting of an experiment becomes extremely difficult as a result of which there may be failures in successful completion of an experiment, technique or study.

The other challenge is that several vaccines and immunostimulants that are developed by biotechnology are known to extremely expensive and it becomes difficult for several people and different countries to afford these vaccines. The costs involved in development of a disease resistant vaccine is known to be a major reason behind several countries not implementing biotechnology methods into their aquaculture.

There are several issues related to the application of biotechnology to aquaculture as it is known to trigger several ethical concerns related to the manipulation of genes for providing food to a major group of population. There are difficulties with respect to the safety of the use of biotechnology principles and perception of the people regarding the utilization of genetic engineering and biotechnology in the development of aquaculture.

There may be several other factors such as intellectual property and accessibility that would affect the growth and development of biotechnology and utilization of these principles in aquaculture.

7.5. CONCLUSION

The application of the techniques of biotechnology in aquaculture is making enormous contributions to the development of aquaculture. There is need to carefully utilize the techniques of biotechnology for the betterment of aquaculture and it is essential to utilize the techniques of biotechnology for the betterment of fish and aquaculture along with the traditional methods of aquaculture.

There is need to make optimum utilization of biotechnology in aquaculture and there is need to take care that the use of biotechnology is mostly need driven and not technology driven.

Biotechnology has promising application in the betterment of quality and productivity of the aquaculture and it is known to play an important role in enhancement of fish production.

The use of biotechnological techniques and principles is known to revolutionize the field of aquaculture and there are significant changes that are associated with the implementation of biotechnology to aquaculture. The use of biotechnology in aquaculture has several concerns but the major advantage is that it ensures food security to the world and improves food availability to the people. The major advantage of the biotechnology application to aquaculture is to provide food to large groups of population by increasing the yield and improving the productivity.

REFERENCES

1. Danish, M., Narayan Trivedi, R., Kanyal, P., Khati, A., & Agrawal, R. (2017). *Importance of Biotechnology in Fish Farming System: An Overview.* [online] Available at: https://www.researchgate.net/publication/323676554_Importance_of_biotechnology_in_fish_farming_system_an_overview [Accessed 30 Jul. 2019].

2. Fao.org. (2019). *Biotechnology in Animal Health and Production.* [online] Available at: http://www.fao.org/3/V4845E/V4845E08.htm [Accessed 30 Jul. 2019].

3. Ictsd.org. (2015). *Strengthening the Global Trade System.* [online] Available at: https://www.ictsd.org/sites/default/files/research/E15_Fisheries_Asche_FINAL.pdf [Accessed 30 Jul. 2019].

4. Imedpub.com. (n.d.). *Insights in Aquaculture and Biotechnology.* [online] Available at: http://www.imedpub.com/insights-in-aquaculture-and-biotechnology/ [Accessed 30 Jul. 2019].

5. Kumar Biswas, J., & Maurye, P. (2017). *Aquaculture Biotechnology: Prospects and challenges.* [online] www.researchgate.net. Available at: https://www.researchgate.net/publication/313839802_Aquaculture_Biotechnology_Prospects_and_challenges [Accessed 30 Jul. 2019].

6. Lakra, W., & Ayyappan, S. (n.d.). *Recent Advances in Biotechnology Applications to Aquaculture.* [online] www.ajas.info/. Available at: https://www.ajas.info/upload/pdf/16_69.pdf [Accessed 30 Jul. 2019].

7. Mayekar, T., Salgaonkar, A., Koli, J., R. Patil, P., Chaudhari, A., Pawar, N., Kamble, S., Giri, A., G. Phadke, G., & Kapse, P. (n.d.). *Biotechnology and Its Applications in Aquaculture and Fisheries.* [online] Aquafind.com. Available at: http://aquafind.com/articles/Aquaculture_Biotechnology.php [Accessed 30 Jul. 2019].

8. dun, O.M., & Uka, A., (2019). *Biotechnology in Aquaculture: Prospects and Challenges.* [online] Ajol.info. Available at: https://www.ajol.info/index.php/njb/article/viewFile/106792/96708 [Accessed 30 Jul. 2019].

8

Social and Economic Aspects of Rural Aquaculture

CONTENTS

9

Issues and Opportunities in Rural Aquaculture

CONTENTS

Aquaculture is an important sector in several countries across the world because it plays an important role in alleviation of poverty and improvement of economic conditions of a country. Aquaculture has special importance because it is known to improve the food availability and establish food security. There are several issues and concerns that are related to aquaculture such as habitat destruction, pollution, invasive species and several others which have been explained in detail. There are opportunities that are being provided by nations across the country and some of the important opportunities related to the aquaculture have been elucidated.

9.1. INTRODUCTION

Aquaculture is known to one of the rapidly growing sectors for the production of food across the world. Aquaculture is popularly known has a sector which can reduce poverty, improve revenue generation, and establish food security.

There are several issues and concerns that are related to the development of aquaculture methods as there are several adverse impacts on the environmental practices and has negative socio-economic impacts. There is need to understand the several issues and concerns that are related to the development of aquaculture because the understanding of issues related to aquaculture might play an important role in improving the areas that are challenging and improve the aquaculture practice methods across the world.

Figure 9.1: Representation of aquaculture.

Source: https://encrypted-tbn0.gstatic.com/images?q=tbn:ANd9GcR6HJwPfH GkFuwgAtzLPonzc2TCbaeDyL6Rvlgle92Yemc1KjUJ

The presence of several issues and challenges in the development of rural aquaculture indicate the need for better understanding of problems related to sectoral sustainability and design ways to make sure that the aquaculture sector may provide fair and important opportunities for the development of human beings. There are several issues and problems in relation to sustainability and enhancement of aquaculture sector to alleviate poverty, establish food security and improve the revenue of a nation.

Aquaculture is an important sector as it provides good quality protein. There a number of people who are dependent on aquaculture sector as it provides a number of employment opportunities, generates revenue, and improves global trade and foreign exchange to a major chunk of population across the world.

Aquaculture is seen as an important method to improve the rural areas of a nation as it plays as an important tool to improve the opportunities in inland and coastal areas. Aquaculture is known to be important in exports to improve the revenue and foreign trade of a nation.

Aquaculture is known to be an export-based industry and has a lot of commercial importance as it gets a lot of employment opportunities and is one of the important sources of revenue for a nation. Aquaculture is known to improve the international trade of a nation because of which it can get a lot of income to a nation by foreign trade. These aqua cultural opportunities are known to improve the livelihood of several people and become an advantage to the poor by providing employment too many people, supplying quality protein at economical prices to poor people and facilitate in establishing food security and improve the lives of several poor people.

The improvement of subsectors of aquaculture such as fisheries and other culture techniques is known to provide improved opportunities and resources to several poor people who get benefited by better practice of growing fish and using improved methods in the development of fisheries.

The use of better methods for the growth of fisheries in a place is known to be extremely important as implementation of novel methods is known to improve the growth and supply of fish from both marine and freshwater fisheries. The fisheries are known to be an important way for several people living in rural areas to improve their revenue.

One of the major challenges that is present in improving the aquaculture practices across the world and especially in rural areas is to provide training programs, awareness in farmers and implementation of better methods in aquaculture systems.

which is achieved mostly by involving all the people who are related to the aquaculture sector. Therefore, the involvement of all the stakeholder's different fields of aquaculture such as planning, decision making, development and management is instrumental for the growth of the aquaculture sector.

- There is a need to create awareness among people to gain access to different resources and utilize them properly in order to develop the aquaculture practice of the place.

- The setting up of aquaculture farm and maintenance of the farm is known to need a lot of money because there is need to create awareness among people regarding different issues related to finances, sources for money, different money lenders and several other issues. If there is awareness regarding these aspects, then people would take financial decisions that are beneficial and lead to the growth of the aquaculture in a particular area.

- Improve the access of people to information or knowledge regarding aquaculture practices and emerging techniques.

- Incorporate and combine the development of aquaculture as a part of coastal and inland watershed management plants.

- Proper planning and better management of resources among several people in order to improve the aquaculture practices of a place.

- Effectively combine aquaculture with the development programs in a country that focus on alleviating poverty and focus on improvement of the poor of the nation. This type of integration of aquaculture development with poverty alleviation programs would improve the employment opportunities and revenue from the aquaculture.

- Manage correctly the investments from both private and public sectors to aquaculture. There is need to focus on development of aquaculture at a commercial level and industrial level in order to promote the aquaculture sector. This can be done by promoting private sector and public sector investments in this area in order to improve the aquaculture sector as a whole.

- Supply different products that are produced by aquaculture according to the needs of the consumer. There is need to improve the efforts and complement the efforts of several areas of food production in aquaculture sectors.

- Improve the collaboration among different stakeholders who are belonging to different countries and regions to facilitate the process of business and development in aquaculture techniques in the field of aquaculture.

- Promote investments from both public and private sectors to improve the aquaculture development.

- The improvement of funds and aquaculture practices in commercial and industrial levels would lead to sustainable development in the aquaculture sector.

9.3. IMPORTANT ISSUES AND CONCERNS RELATED TO RURAL AQUACULTURE

There is increase in population across the world. In this way, there is an increase in the need for food and healthy protein across the world.

There are different techniques that are used in aquaculture that are being used to improve the yield of a particular variety, and as a result there is some amount of impact on the environment because of the aquaculture practices. There are many issues that are related to the aquaculture sector and some themes are mentioned in the following subsections.

9.3.1. Impacts of Forage Fish

Aquaculture generally involves feeding the aquatic species with vegetarian food but in the recent times there are different species of aquatic organisms such as salmon and others which may need wild fish as a source of fish meal and need fish oil for their food. This indicates to the fact that there is need for a kilogram of wild fish that is needed for the growth of a kilogram of farmed fish. This discovery regarding the culture of fish was published in a paper that was published in the journal nature.

relation to shrimp culture because of making mangrove cutting illegal in several countries.

There was a new disease that attacked shrimps and led to huge losses because of this disease Early Mortality Syndrome, which arose in China and that, had spread throughout Southeast Asia. This Early mortality syndrome was responsible for the death of different types of shrimps and the loses were estimated to be nearly $1 billion because of this Early Mortality Syndrome (EMS).

9.3.4. Aquaculture and The Environment

Several fish species can be cultures into tanks that resemble an aquarium. These would contain huge volumes of dirty water that needs to be changes. There are chances for large amounts of feces, nutrients and chemicals that might be released into the environment which might create pollution in the environment. The release of these wastes, feces, nutrients, and chemicals into the environment may result in algae blooms that would subsequently remove the dissolved oxygen in water leading to eutrophication. If there is very less or no oxygen, then there are chances for the fish to die in the environment.

There may different types of chemicals such as antibiotics and different types of water treatment agents which might use as a part of aquaculture which would create a lot of pollution. There is need to treat the wastewater before the discharge of water into the open lands to avoid contamination.

9.3.5. Aquaculture and Disease

The growth of different aquatic organisms such as fish, shrimps and others may lead to the spread of several disease-causing parasites in the world. The studies reveal that there is need for good maintenance of the commercial chicken coops to keep them disease free, similarly there is need for notorious spread of disease in farmed fish and shell wish when they are grown in aquaculture ponds.

There may be reduced disease resistance in farmed which because of which they would easily catch a disease. The chances of catching several parasites such as sea lice and other parasites is more in farmed fish when compared to the fish that grown and live in a natural environment. The farmed fish have a greater risk of disease when they are exposed to unprocessed fish as feed rather than safer processed fish pellets.

9.3.6. Invasive Species in Aquaculture

Aquaculture is one of major reasons for the arrival of several foreign invasive species to come up in a new area. There is chance for the occurrence of several invasive species to occur under the right conditions and which may affect the aquaculture of a particular place in several ways.

The farmed fish may escape from their pens and might damage the environment by the threatening several species of native fish populations ad they may carry several disease and parasites along with them and pass the diseases to several organisms. The parasites that are transferred from one species to the other might affect the native species and kill this species.

Figure 9.3: Representation of invasive species in aquaculture.

Source: https://www.dec.ny.gov/images/fish_marine_images/rapa.jpg

This would lead to the escaped farm fish to compete with other organisms for their food and habitat and lead to the displacement of indigenous species. These farm species might enter the ecosystem of native fishes and compete with the native fishes for food and habitat.

There may be several native farm fish that might breed with the wild native fish and they may weaken the natural gene pool by threatening the long-term survival and lead to the evolution of several wild species with different gene pool. The impact of this novel fishes on the environment may not be known and sometime could be extremely dangerous.

9.3.7. Secondary Impacts of Aquaculture

There are several different secondary impacts that may occur in the natural environment because of aquaculture and some of them may be because of species of farmed fish might need an alternative food source for their healthy growth and development.

Therefore, to suffice the needs of farmed fish there may be several native and wild species of fish that were over fished which is a threat to different natural sources of fish. There are several species of farm fish, which are carnivorous by their food choice, and these fish are mostly fed by whole fish or pellets that are made from the fish. There could be several species of fish such as mackerel, whiting and herring which are mostly threatened and are getting endangered because they are often used as feed for several farmed fish.

9.3.8. Effects of Construction on Aquaculture

There are chances for both land and aquatic animals to lose their natural habitats while building aquaculture farms along the coastal line. If there is construction that is happening, then the clean and natural water can be accessed for their processes, and one such example is the chopping of mangrove forests in Asia and Latin America to construct the shrimp farms.

9.3.9. Lack of Trained Personnel

There is one major constraint and limitation with respect to the growth of aquaculture which is common to several countries and that is known to be the lack of proper personnel that can work in this sector. There are very few limited workers who are working in this sector and most of them lack proper training because of which implementation of novel techniques and methods in the production of fish is known to become extremely difficult.

There is a lot of focus and interest for people to do jobs related to administration and research work in aquaculture and the work associated with the growth of aquatic species is majorly done by those people who work in agriculture, water, or forest departments. These people lack basic training in aquaculture and may not be aware of the methods which can improve the production of aquatic species in aquaculture and this is a major drawback for promotion of aquaculture in a nation. There is need for special extensive training programs that must be conducted to several personnel for the improvement of aquaculture in the region.

9.4. IMPORTANT CONSTRAINTS AND CHALLENGES IN AQUACULTURE

There are two major constraints that are present in relation to issues, concerns and problems related to the development of aquaculture and they are as follows:

- There is a lot of inconsistency that is present in the objectives that are related to the evaluation and development of several aquaculture programs.

- The second constraint regarding aquaculture that is present in different countries is mostly due to the lack of properly trained people to undertake aquaculture programs.

Aquaculture is considered as one of the most important ways to produce fish that can be eaten by different people in rural areas. Aquaculture is capable of establishing food security to a variety of people present across the world and as a result of which there are several governments who are supporting the development of aquaculture. There are several fish production programs that are considered as economical methods for the production of fish and they are widely supported by governments of several countries as they are known to generate a lot of revenues and monetary benefits to the local population.

Aquaculture is known to be one of the economically sustainable industry as it is known to generate a lot of revenue to the country and can make enormous contributions to the food production of a country along with providing huge number of employment opportunities to the youth of the nation. There are several different ideas and objective of people which are responsible for mixed reviews and results in relation to the field of aquaculture.

There is need to understand that there is no difference between conventional fishing and aquaculture because of which aquaculture is being developed in those areas and countries which have a decline in fish production. There are several measures and steps that are being taken to improve the growth and development of fisheries in these countries which see a rapid decline in the production of fish.

9.5. OPPORTUNITIES IN AQUACULTURE

The aquaculture sector is known to be an important sector that is responsible for the growth and development of aquaculture in nation. Aquaculture is known to create a number of employment opportunities and alleviate the poverty of a nation. There are different opportunities that are present for aquaculture in any nation and they are as follows:

- Several communities regionally and nationally which have recognized rural aquaculture as a major food production system and is known to provide ample food protein to different classes of people at affordable prices.

- The growth of the aquaculture sector is largely dependent on the availability of trained professional which is achieved by conducting several workshops and training programs to update people regarding the latest advancements and techniques that can be used for the improvement in rural aquaculture.

- Several places which has good soils and climatic conditions that are present across several countries in the world that would facilitate the growth of aquaculture.

- Various countries which are making good policies and legal framework to support the growth of aquaculture.

- Some agricultural farmers who are being encouraged and being trained to pursue aquaculture practices.

- Department of fisheries which is being setup in several countries to support the aqua culture farmers for the improvement of aquaculture sector.

- The development of aquaculture is a part of the national development plans in several countries.

- Aquaculture is being included easily into several small-scale farming practices and is being focused for the improvement of aquaculture.

9.6. CONCLUSION

There were several failures with respect to the growth and development of aquaculture in several countries from the colonial period because of which there is need for several nations to take up aquaculture as a part of their national development plan and must work on development of aquaculture as a high priority in order to encounter development of aquaculture. There are different countries which have combined aquaculture as a part of rural development plans in order to improve the aquaculture in rural areas.

There is need to take care regarding the different species that are being grown as a part of aquaculture. There is need to take healthy species which are disease resistant and disease free in order to see financial losses. The choice of species is an important step in aqua culture as there is special cultural importance to different types of aquatic organisms present in different areas. There is a lack of extension service in many nations which is a major drawback in these countries.

The lack of proper personnel who are trained in the area along with different negative impacts of aquaculture on environment and other implications of the aquaculture must be considered for the improvement of the aquaculture practices of a country. There is need to be aware of the opportunities and issues related to aquaculture in order to facilitate the growth of aquaculture in a country.

9.7. CASE STUDY: AGRICULTURAL AND RURAL DE-VELOPMENT/FISHERIES

Since rising sharply in 2008, the price of food has hit peaks again in 2011 and 2012. Even after these three peaks, food prices have continued to rise. These soaring prices pose a threat to the food security of developing nations. They are particularly damaging to the urban poor and to the rural poor; such as small-scale farmers and fishers who cannot even produce and catch enough food to meet their own needs.

JICA is providing cooperation to address the issues of agricultural, maritime and rural development. The goal is to contribute to the Millennium Development Goal (MDG) of "eradicating extreme poverty and hunger" and its successive agenda Sustainable Development Goals by offering aid for food production, food supply and nutrition to the residents of both rural and urban areas.

9.7.1. Introduction

The environment surrounding agricultural and rural development has been diversifying because of such factors as the rapid advance of globalization, climate change, skyrocketing food prices, growing demand for biofuels, changing food preferences as personal incomes rise, the expanding participation of the private sector, global competition for farmland, and post-conflict rehabilitation.

As in many developing countries, farmers account for the majority of the population and three-fourths of impoverished people live in rural areas, rural residents in developing countries are greatly affected by these changes.

9.7.2. Stable Food Supply

According to an estimate by the Food and Agriculture Organization of the United Nations (FAO), the number of people in developing countries suffering from malnutrition is expected to be around 805 million during the period from 2012 to 2014, remaining at a high level. It will therefore be difficult to achieve one of the targets of Goal 1 of the MDGs, namely to "Halve, between 1990 and 2015, the proportion of people who suffer from hunger."

Consistently providing people with the food that they need (food security) is the foundation for economic and social stability and an important policy issue. However, the food security of many developing countries is easily affected by due to such factors as insufficient capacity of government in planning and implementation, underdeveloped agricultural infrastructure, low levels of production technology, and inadequate distribution systems that threaten the food security of citizens.

Consequently, this situation results in health deterioration, causes the outflow of valuable foreign currency to pay for food imports, and accelerates urban shifts as well as the abandonment of farming. These influences in turn lead to a worsening of poverty in urban areas, a contributory factor to social and economic instability in a country.

9.7.3. Reducing Rural Poverty

According to the World Bank's 2013 report, the number of people living in extreme poverty (on less than \$1.25 a day) dropped remarkably over the last 30 years. In 1981, half the population of developing countries lived in extreme poverty; this rate dropped to 21%. However, this reduction owes much to the economic development of East Asia, and in real numbers, 1.2 billion people still live in extreme poverty. Currently, one-third of these extremely poor people live in Sub-Saharan Africa.

On the other hand, while self-sufficiency rates for major grains have improved somewhat and the urban middle class is growing in such regions as South America and Southeast Asia, urban and rural disparities exist. In these regions, there is a need to devise ways to narrow the increasing economic gap between urban and rural areas. It has been reported that the poverty reduction effect of growth driven by agriculture is at least twice that of growth driven by other industries.

Moreover, in countries with lower income levels, such as in Sub-Saharan Africa, the high proportion of agriculture in the GDP means that in many cases people are pinning their hopes on agriculture as the source of economic growth.

9.7.4. JICA Activities

JICA's cooperation in agricultural and rural development aims to ensure a stable food supply to people in both rural and urban areas and reduce poverty in rural communities—thereby driving economic development at national and regional levels. Through these activities JICA strives to contribute to achieving goals and targets of the MDGs and SDGs. For this reason, JICA has established the following three specific cooperation objectives.

1. Sustainable Agricultural Production

Risks involving the food supply are a complex combination of short-term and long-term factors. Short-term risk factors include poor harvests owing to bad weather and accompanying speculation. Long-term factors involve population growth in emerging countries, changes in the demand structure in those countries, limitations on production resources such as land and water, vulnerability to climate change, and competition between rising demand for biofuel and food.

As a result, dealing with these issues requires determining measures for each cause based on the differing circumstances of each region. JICA is aiming to achieve sustainable agricultural production in order to address these diverse problems. In its approach to enable stable agricultural production, first, JICA provides aid for drafting agricultural policies reflecting the characteristics of the partner country's overall agricultural sector. Based on these policies, JICA provides cooperation from the perspective of the overall value chain, from production to distribution and sales.

Initiatives include establishing, maintaining, and managing infrastructures for agricultural production such as irrigation systems; improving the procurement and use of seeds, fertilizer and other agricultural production materials; and establishing and utilizing production technology for grain, livestock and other items while supporting institutional strengthening of associated organizations. In addition, JICA is taking action regarding increasing the resilience of agriculture to climate change.

Activities include facilitating sustainable land use, development and study on appropriate technology, developing second-generation biomass energy that does not compete with food production, establishing stockpiling systems, using agricultural statistics and introducing weather insurance, and promoting the private sector's entry into the market. For example, in Myanmar, irrigation systems have been maintained by ODA Loan to improve productivity and profitability of farmers.

At the same time, JICA is preparing comprehensive cooperative projects, including developing policies and systems for the promotion of irrigated agriculture, improving production technology of major crops using irrigation water, properly introducing and handling agricultural machines and materials, and facilitating cooperation with the private sector. Furthermore, along with their rising incomes, citizens of developing countries are increasingly demanding high value added agricultural and livestock products as well as taking a greater interest in such food issues as quality and safety. These issues also need to be addressed.

2. Stable Food Supply

Sustainable production is the premise for the provision of a stable food supply to the people of a country. In addition, ensuring a stable supply requires the establishment of food supply and demand policies for an entire country that reflect international food security. Creation of a framework for food imports and the proper use of food aid are also necessary.

Africa accounts for the largest portion of people suffering from malnutrition in the world (35% of the total population in 2011), and is in great need of expanded food production. The amount of rice consumed in Africa is growing rapidly and there are excellent prospects for achieving sustainable growth in rice production. Therefore, rice is believed to be the key to eradicating the lack of food security on the continent.

With other donors, JICA launched an initiative called the Coalition for African Rice Development (CARD) in 2008. In order to contribute to food security, the goal is to double rice production in Africa from 14 million tons to 28 million tons over the 10-year period ending in 2018.

To reach this target, JICA is providing aid for the formulation of National Rice Development Strategies in the 23 rice-producing countries in Africa and for boosting rice production in line with the strategy of each country. As for the entire Sub-Saharan Africa region including CARD member countries, rice production increased 59 % from 14 million tons in the reference year to 22.23 million tons in 2013.

3. Promoting Dynamic Rural Communities

For rural development that reduces poverty, it is important to aim for social changes and invigoration in rural villages from the standpoint of developing agricultural economies and enhancing the livelihood of people. Accomplishing this goal requires going beyond simply raising productivity. For instance, the distribution and sale of food must be improved, the food processing sector energized, export promotion measures strengthened, and agricultural management must also be upgraded to increase non-agricultural income and such.

Furthermore, aid is needed that brings together a diverse range of fields. Local administrative functions must be strengthened and rural infrastructures such as community roads and drinking water supplies established. The rural living environment must be improved and level of health and education for residents enhanced.

Other examples of aid are the participatory development of rural areas and narrowing gender gap. Moreover, for post-conflict countries, because agricultural and rural development is often a key component of aid, JICA gives priority to these activities. To stimulate rural development, JICA supplies aid to local administrative institutions in drafting development plans with the participation of rural residents. JICA also provides aid for the establishment of implementation systems that enable the community to raise income and improve people's livelihood, through improving the processing, distribution and sale of agricultural products.

For example, in the technical cooperation projects implemented in Kenya to support improvement of smallholder farmers' livelihoods, the Smallholder Horticulture Empowerment and Promotion Project (SHEP, 2006–2009) and the following Smallholder Horticulture Empowerment and Promotion Unit Project (SHEP UP, 2010–2015) have supported the farmer groups to change their attitudes from "grow and sell" to "grow to sell," introducing the concept of market-oriented farming.

As a result of various support activities—the SHEP Approach—to make farmers manage market-oriented agriculture by themselves, the horticultural incomes of the farm households involved in the projects have increased. The effectiveness of the SHEP approach has been recognized by other donors, including the United States Agency for International Development (USAID), and Japanese Prime Minister Shinzo Abe also touched on it at the opening session of the Fifth Tokyo International Conference on African Development (TICAD V) in 2013. In response to this, JICA places priority on implementing the SHEP approach in other African countries, and it is being implemented in 18 countries as of June 2015. Meanwhile, the third phase of the SHEP Project, the Smallholder Horticulture Empowerment and Promotion for Local and Upscaling (SHEP PLUS, 2015–2020), started in Kenya in March 2015. SHEP PLUS works on establishing implementation mechanisms for further promoting SHEP approach in Kenya, where a devolution process has taken place, through reflecting the lessons learned by supporting other African countries.

9.7.5. Fisheries

9.7.5.1. Overview of the Issue

Fisheries resources from the oceans, rivers and lakes are important sources of food for people in developing countries. According to FAO, fishery products constitute nearly 20% of animal protein intake in developing countries and they are often among limited choices of affordable protein source. As such, the fisheries sector plays an important role in terms of providing a valuable means of livelihood for most vulnerable population such as women-headed households and those people who do not possess production assets. Developing countries account for 54% of the world's exports of fisheries products in value terms and 60% in volume terms (estimated live fish weight in 2012).

These rates have been increasing for the past 10 years, making this industry vital to the economies of developing countries. World fisheries and aquaculture production is currently 158 million tons (as of 2012). However, the capture production from marine waters reached a peak in the 1990s and it is believed that these resources have been almost fully utilized since then. In recent years, the stagnant capture production has been supplemented by rapidly growing aquaculture production, which now accounts for 40% of total fisheries production.

Even in the developing countries the decline in fishery resources is evident, probably due to overfishing and the destruction of the natural environment and ecosystems. However, the practice of proper management of fishery resources has not yet been sufficiently adopted by fishers. Fishers are often deprived of alternative means of livelihood and hence they have a strong tendency to prioritize immediate economic returns rather than long-term sustainable benefits. Therefore, implementing effective management of fisheries resources, which can be accepted by a majority of fishers, is a key challenge in these countries.

9.7.6. JICA Activities: Fisheries

As mentioned above, the lack of proper management of fisheries and deterioration of environment, which causes further decrease in valuable fisheries resources, is a major issue in the fishery sector. Fishing villages, which are often located in rural marginalized areas, also face a wide range of social issues including chronic poverty.

It is hence fundamental that fisheries management issues are addressed within the overarching framework of "fishing communities (villages) development," which adequately incorporates the aspect of the livelihoods of all members of the fishing community. JICA's cooperation in the fishery sector has three main objectives:

- ensure the stable supply of food to local people,
- eliminate malnutrition by providing valuable nutrition, and
- reduce poverty by providing a means of earning livelihoods to the poor.

Adequate management of fisheries resources will provide a good basis for achieving these objectives and is a key to fishing village development. JICA has set the following three pillars for its cooperation.

1. Vitality in Local Fishing Communities

Empowering fishing communities to alleviate chronic poverty requires a comprehensive approach. The efforts to promote sustainable fisheries resource management would be more effective if these are supplemented by activities that stabilize communities' livelihoods. These may include the promotion of alternative income generation activities such as agriculture, and the provision of education, health services and other social development programs.

JICA provides support for fisheries infrastructure development such as construction of landing ports and markets that promote efficient fish marketing as well as community members' collective actions. JICA also works to improve the capacity of fisheries organizations as well as women's group fish processing and sales activities.

2. Stable Food Supply (Effective Utilization of Fisheries Resources)

Food security issues are getting more serious in a number of developing countries due mainly to rapidly growing population. This situation inevitably puts further pressure on utilization of fisheries resources. In order to ensure sustainable supply of fish while avoiding overexploitation of resources, the fisheries sector urgently needs to make a major shift in production efforts; from "fishing" to "fish farming." In response to such needs, JICA is now exerting efforts on promoting fish farming. In promotion of aquaculture in rural areas, JICA takes a unique approach of "farmer-to-farmer training." In addition, JICA is assisting human resources development in aquaculture, targeting researchers, technicians and extension workers.

Fish and fishery products are highly perishable. In developing countries where the distribution infrastructure is underdeveloped, post-harvest loss is an issue. JICA provides support for improvement of related facilities of fish landing, distribution and marketing for better quality and hygienic control. Such efforts are complemented by technical assistance on fish processing and preservation.

3. Appropriate Preservation and Management of Fisheries Resources

Fisheries resources are basically "renewable" resources if proper management is put in place. Having learned the lesson that government led top-down approaches may not be an effective way to promote fisheries resource management, JICA is applying a co-management approach, incorporating awareness building and capacity development among groups of fishers and facilitating collaboration mechanisms among key stakeholders.

JICA also put its emphasis on capacity development of fisheries administration for supportive policy framework and effective implementation. Collection of scientific data for informed decision-making and support for regional initiatives are also a part of this direction. With the participation of local fishers, JICA works to preserve and restore critical habitats of marine and inland water ecosystems, including seagrass beds.

REFERENCES

1.   Albert, N., & Simbotwe, M. (2014). *Challenges and Emerging Opportunities associated with Aquaculture development in Zambia.* [online] Fisheriesjournal.com. Available at: http://www.fisheriesjournal.com/archives/2014/vol2issue2/PartE/53.pdf [Accessed 29 August 2019].

2.   Fao.org. (2019). *2. Problems and Prospects of Aquaculture Development in Africa.* [online] Available at: http://www.fao.org/3/H2920B/h2920b02.htm#2.%20problems%20and%20prospects%20of%20aquaculture%20development%20in%20africa [Accessed 29 August 2019].

3.   Greenberg, P. (2014). *Environmental Problems of Aquaculture.* [online] Earth Journalism Network. Available at: https://earthjournalism.net/resources/environmental-problems-of-aquaculture [Accessed 29 August 2019].

4.   Hussan, A., Gon Choudhury, T., K Gupta, S., & Vinay, T.N. (2016). *Common Problems in Aquaculture and Their Preventive Measures.* [online] Available at: https://www.researchgate.net/publication/312947314_Common_problems_in_aquaculture_and_their_preventive_measures [Accessed 29 August 2019].

5.   JICA Annual Report 2015 (2015). *Agricultural and Rural Development/ Fisheries.* [ebook] Available at: https://www.jica.go.jp/english/publications/reports/annual/2015/c8h0vm00009q82bm-att/2015_35.pdf [Accessed 30 Jul. 2019].

6.   Quality, S. (2019). *Soils and Water Quality | NC State Extension Publications.* [online] Content.ces.ncsu.edu. Available at: https://content.ces.ncsu.edu/soils-and-water-quality [Accessed 29 August 2019].

7.   Souza, M. (2019). *Five Problems Inherent to Aquaculture.* [online] The Balance. Available at: https://www.thebalance.com/aquaculture-problems-inherent-to-aquaculture-1301970 [Accessed 29 August 2019].

8.   Souza, M. (2019). *Find Out Where Your Seafood Comes From.* [online] The Balance. Available at: https://www.thebalance.com/top-aquaculture-countries-1301739 [Accessed 29 August 2019].

10

Future Aspects of Rural Aquaculture

CONTENTS

With the passage of time, there has been development a new global scenario of rural aquaculture. The future of the rural aquaculture totally depends upon the practices that are being put to use in the modern times. Along with the practices, it is also important to focus on the strategies and the possibilities of maintaining the sustainability of the productive chain. The usage of techniques that are equipped with the modern technology will play a significant role in the development of the proper rural aquaculture. Thus, it is very important to understand and observe various factors such as technology, global demands and various economical aspects that can affect the in the near future. Although, there are several practices that will be put to use in the future. The main aim of these practices will be to develop certain level of sustainability in the field of rural aquaculture in the upcoming future.

10.1. INTRODUCTION

The field of aquaculture has a significant role in rural development. Three quarters of aquaculture production comes from the low-income nations, the most important region being Asia, in this region Chinese production predominates. In some of the developing countries, it has been noticed that in the modern times, rural aquaculture supplies near about 20% of the domestic fish consumption.

Integrating aquaculture into the rural economy can bring several numbers of advantages, as well as environmental and social dangers, particularly in coastal areas. Lessons must be learnt from the case of uncontrolled extension of intensive marine shrimp production. To be particular, rural aquaculture has great potential to enhance the level of food security and be sustainable, especially in the environmental aspect.

In the developing economies, peoples' livelihoods, which comprises of aquaculture, benefit from participatory methods, which build management capacity. In case of inland areas, fry nursing networks signify low-risk entry points for rural development, and fish-in-rice systems consists of wide range of application. On the other hand, in case of coastal areas, reforestation can provide advantages to coastal defenses and aquatic resource production, while integrated pond-dyke cropping systems in delta regions have established complementary resource and energy flows.

In more of the developed nations, where the purpose or aim is the development of remote rural economies, the stability and environmental influence of aquaculture should be the main considerations in the upcoming future planning.

Figure 10.1: With the advancement of technology, it has been noticed that rural aquaculture has also evolved.

Source: https://www.nps.gov/subjects/culturallandscapes/images/36990359233_b2256694a7_k.jpg?maxwidth=1200&maxheight=1200&autorotate=false

Aquaculture has been considered as one of the most rapid and technically innovative of food production segments all over the world, with important investment, scientific and technical development and growth of production in many of the regions of the world over the time period of previous twenty years. While this has had an important influence on the supply of aquatic food products all over the world and had a significant effect in rural and urban food supply and also it helps in offering employment services in many of the developing nations, growth and growing internationalization has not been without concern for the usage of the natural resource, environmental effect and social disruption.

In the present times, the expectations for production and diversification are considered as significant and though the scientific and technical means are already available in order to meet much of the planned targets, practical constraints of investment, profitability, resource accessibility and system effectiveness are likely to become far more significant constraints for the upcoming or future times.

There are some studies and reports that provides a contemporary point of view on the means in which the department might develop, its interactions with constraints and the approaches that may be needed in order to ensure that future development is both positive as well as sustainable.

10.2. FUTURE OF AQUACULTURE: SUSTAINABLE AQUACULTURE

The efficient rural development comes through sound governance, involvement at all shareholder levels, people-focused incorporated sustainable development and a multi-sectoral agenda. Policy coherence must be considered as a main purpose, developed by the means of wide-ranging public participation and, where requires, through the promotion of efficient representative associations. Various types of aquaculture form a significant component within the agricultural and farming systems development.

Figure 10.2: There are many practices that are focus on the development of sustainable rural aquaculture.

Source: https://live.staticflickr.com/4014/4478993436_c501b6dc0a_z.jpg

Much greater focus is laid on advocacy (exterior to the subsector) is needed in order to increase the awareness of the role for aquaculture in rural development and to increase the stakes for institutional alteration.

The main purpose of regulation and policy should be to internalize the external impacts of aquaculture (for instance, the "polluter pays" principle). Special emphasize is needed in order to empower and connect stakeholders to policy decisions.

To provide the food products for the increasing population, serious environmental issues may result. In spite of many advantages from the green, blue, and silver revolutions that were adopted in India, there has been much concern that were resulting from the intensive agricultural practices that were carried out, this gave rise to environmental issues in both terrestrial as well as aquatic ecosystems.

The growing demand for aquatic resources also caused inland fisheries to reduce over the previous few decades. The site of the aquaculture projects, devastation of landscape, soil and water pollution by the waste of the pond effluents, over-exploitation of the significant fish stocks, exhaustion in biodiversity, fights over agriculture and aquaculture between several groups of stakeholder over resource and space allocation, and international fish trade controversies have endangered the long-term sustainability of the industries in aquaculture and fisheries.

The subject of sustainable aquaculture has not been sufficiently projected in the context of the existing aquaculture practices that are targeting to raise the level of a rural economy. There are many reports and studies that briefly defined the main concerns of aquaculture unsustainability in the context of intensive aquaculture, nutrient enrichment syndrome, soil and groundwater salinization, devastation of mangroves, loss of biodiversity, marine pollution and loss of fish stock, application of aqua chemicals and therapeutics, hormone remains, and many more.

The approached or strategies for sustainability have been outlined with respect to rice-cum-fish culture, carp polyculture, incorporated farming with livestock, rural aquaculture, establishment of small farms, wastewater-fed aquaculture, crop rotation, probiotics, quality of feed, socioeconomic considerations, ecological regulations and fisheries acts, transboundary aquatic ecosystems, effect of alien species, ethical impacts of intensive aquaculture, responsible fisheries, and environmental impact assessment.

A proposed model underlines the feedback mechanisms for attaining the long-term sustainability through the enhanced farm management practices, integrated farming, use of selective aqua chemicals and probiotics, preservation of natural resources, regulatory mechanism, and policy instruments.

10.3. THREATS AND WEAKNESSES IN FOOD-FISH PRODUCTION

10.3.1. Fish-Feed Extraction

It has been observed that the production of high-value, high trophic food fish continue to attain the greatest support in the political as well as economical manner. The increasing demand and increasing incomes are making an incentive for the farmers to move away from the low-value food fish for the purpose of domestic consumption, to high-value fish for the purpose of export.

The abstraction of small or low-value fish species in the wild, also called as 'trash' fish, is considered as an important component of aquaculture as many of the high-value food fish are carnivorous.

The abstraction of wild fish for the purpose of non-human consumption (for instance, fishmeal, fish oil and feed for farmed fish) is progressively being questioned or interrogated for its lack of sustainability and usage of resources.

The capture of small fish for the production of feed is difficult in upholding the balanced marine ecosystems and food webs for the greater predators. With 70% of ocean fisheries in the requirement of urgent management and 50% of these are already fully exploited, the utilization of the captured fish species for the purpose of fish farming is likely to initiate some kind of conflict in the near future.

The production of fishmeal and fish oil needs near about thirty 30 million tons of captured fish every single year. Feed is manufactures predominantly from small pelagic marine fish, comprising of sardines, anchovies, sand eels, herring and mackerels.

Future competition is likely among the human consumption requirements and animal feed needs at current rates of production. The percentage of fish for the purpose of human consumption has increased significantly since the year 1990s, but the utilization of fish feed in aquaculture has also experienced substantial growth over the time period of previous twenty year.

The approaches that are used for capture and storage of these smaller fish makes them inappropriate for the purpose of human consumption, but the continuous growth of the production of aquaculture means it is not sustainable to continue drawing the trash fish from the oceans. Finding a substitute source of feed is considered as an international research priority.

Commercial aquaculture has exceeded the population of poultry and livestock as the main consumer of fish-based animal feeds. With emphasize on high-value fish species, several numbers of producers depend on carnivorous species with high-trophic levels.

As this industry depends on the export market, there is a requirement to decrease the reliance on capture fisheries for obtaining the raw materials for feed. On the other hand, small-scale farming is restricted by the availability and price of feed, restricting the small-hold farmers to low-trophic species, which are more famous in domestic markets.

10.3.2. Disease

Disease and chemical contamination of farmed fish is considered as an ongoing management concerns for farmers. Disease is most frequently caused by the stress factors, induced by the alterations in the environment.

There are many factors that increase stock susceptibility to disease. These factors are the Lowered environmental health, increased levels of waste and contaminants in waterways, high stock density and low-quality. The closed farming systems where the breeding environment is closely supervised, are considered as the safest or secure option for the prevention of disease and water quality control. In addition to it, the closed farming systems avert escape, guarantee the availability of optimal levels of feed and water, and remove any risk of predators. The migration of the live aquatic animals through the boundaries is a main cause of the spread of sicknesses and pathogens inside the aquatic environment. The enhanced technology and management systems are needed in order to regulate the diseases in a controlled manner, enhance the quality of water and guarantee the sustainable production can be increased in order to meet the future needs or demands.

10.3.3. Environmental Degradation

A major issue with the practice of aquaculture is its environmental influence and degradation of water quality from its production procedures. The wastewater from the ponds causing environmental contamination, nutrient build-up (mainly organic nitrogen and phosphorus) and wastes in ecosystems, land clearing and chemical pollution, are just a few of the negative effects if systems are not managed in a proper way.

Chemical pollution and the pollution from wastewater can severely decrease the level of oxygen in water, create algal blooms and destroy corals and other habitats. Antibiotics that are added to fishmeal, or chemicals that are added to pens as a disease preventive, move directly into the water.

The huge densities of the populations of fish in net pens increase the pollutant or contaminant outputs into surrounding waters, putting augmented stress on the marine ecosystem. The estimations show that a salmon farm of 200,000 fish discharges levels of nitrogen, phosphorus and fecal matter equal to the untreated sewage from more than twenty thousand individuals.

Figure 10.3: With the increased activity in the rural aquaculture, there has been a deep impact on the environment as well as the biosphere.

Source: https://upload.wikimedia.org/wikipedia/commons/5/53/Environmental_degradation_at_Kinder_Low_-_geograph.org.uk_-_834056.jpg

Degradation of land and altered river ecology are the causes of inland farming. These are also considered as challenges which are required to be addressed in order to guarantee that production has a minimal impact on natural biodiversity and ecosystems.

Commercial aquaculture poses a specific set of difficulties, with large-scale production and restricted management in some instances which gives rise to serious environmental harm and ecosystem degradation.

10.4. GETTING AQUACULTURE GROWTH RIGHT: FIVE APPROACHES

There many reports that suggested five approaches in order to help get aquaculture growth right:

10.4.1. Capitalize in Technological Innovation and Transfer

Aquaculture is considered as a young industry—decades behind that of livestock farming. The enhancements in breeding technology, disease control, feeds and nutrition, and low-impact production systems are interconnected areas where science can complement traditional knowledge to enhance the effectiveness.

These kinds of innovations, whether led by farmers, investigation institutions, businesses, or governments, have been behind productivity gains in every single part of the world. For instance, in Vietnam, a breakthrough in catfish breeding around the year 2000—complemented by extensive adoption of high-quality pelleted feed—unlocked a boom in growth and intensification of the production. Vietnamese catfish production grew from 50,000 tons in the year 2000 to over one million tons in the year 2010, even though the total catfish pond region of the country only doubled during that period of time.

10.4.1. Focus beyond the Farm

Most of the aquaculture guidelines and certification schemes emphasize at the individual farm level. But having many producers in the similar area can give rise to increasing environmental influence—such as water pollution or fish diseases, even if everybody is following the law.

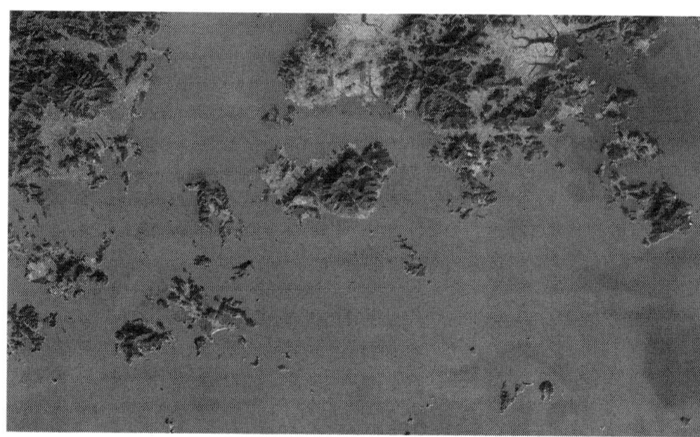

Figure 10.4: It is very important to focus on the land farms along with the rural aquaculture sites in order to maintain the sustainability.

Source: https://live.staticflickr.com/7723/17322757055_f2d02172b1_b.jpg

Spatial planning and zoning can guarantee that the operations of the aquaculture stay inside the surrounding ecosystem's carrying capacity and can also decrease the fights over resource use. Norway's zoning laws, for instance, guarantee that the salmon producers are not excessively concentrated in one area, decreasing the disease risk and helping mitigate environmental effects.

10.4.3. Shift Incentives to Reward Sustainability

A variety of public and private policies can help the farmers in providing them with the incentives to practice more sustainable aquaculture. For instance, the government of Thailand has provided free access of training, water supply, and wastewater treatment to the shrimp farmers who are working legally in aquaculture zones. In addition to it, the government has also provided low-interest loans and tax exemptions to small-scale farmers, assisting them in using improved technology that increases rate of productivity, decreasing the pressure to clear new land.

10.4.4. Leverage the Latest Information Technology

The advancements in satellite and mapping technology, ecological modeling, open data, and connectivity mean that global-level monitoring and planning systems that inspire the development of sustainable aquaculture may be possible in the present times.

A platform that is incorporating these technologies could help the governments in improving the spatial planning and monitoring, help the industry plan for and demonstrate sustainability, and assist civil society report achievement stories and it holds the industry and government responsible for wrong conduct.

10.4.5. Eat Fish That Are Low on the Food Chain

The process of fish farming can ease pressure on marine ecosystems if farmed fish do not require huge amounts of wild fish in their diets. Therefore, the consumers should demand species that feed low on the food chain—"low-trophic" species like tilapia, catfish, carp, and bivalve mollusks.

In the developing economies, where the consumption of low-trophic species is still dominant, focus should be on these species even as billions of people all over the world enter the middle class in the near future.

At the same point of time, because fish are considered as a main source of nutrition for more than a billion deprived or poor people in the developing world, increasing aquaculture in order to meet the food and nutritional required of these consumers will be important.

10.5. DEVELOPMENT OF RURAL AQUACULTURE IN FUTURE

There exist two basic systems for aquaculture, land based, or water based. Land based systems comprises of ponds, rice fields, and RAS systems. On the other hand, the water-based systems make use of the existing water bodies along with cages or other enclosures that are created for containing or appealing the aquatic species. Both lands based or water-based systems offer the upcoming opportunities for new individuals to explore into the aquaculture industry. Water based systems would be particularly useful for landless people to attain the steady income and food safety. Since its initial phase in the region of China, semi-intensive fish culture has been paired with other kinds of agriculture in order to diversify the crops, and therefore reduce the danger of crop loss.

The deprived or poor farmers that already have a pond or rice field on their property can simply stock that water body with small fry that can be attained for comparatively low price. The technology that is used in order to raise fish in these surroundings is of low cost (if is even needed to purchase) and comparatively easy to learn providing the appropriate education.

There were many reports or articles who stated that "addition of fish and other aquatic organisms has been established in order to generate income in rural societies." The advantages of aquaculture for enhancing the living of the poor or deprived people comprises of food safety, protein food source, employment and income.

These systems have been proved to enhance the rate of productivity and profitability of small-scale farms. In the recent times a term "rural aquaculture," having been presented. This is defined as "the farming of aquatic organisms by small-scale households utilizing mainly extensive and semi-intensive husbandry for the purpose of household consumption and/or income." It has been observed that rural aquaculture has turned out to be the fastest growing food-production technology in the world.

There are several opportunities that are being projected for the development of the rural aquaculture to help in poverty alleviation. One such project that is being supervised or led by the Fisheries and Marine Institute of Memorial University is known as the Cambodian Sustainable Rice Fish Integration (SRFI) Project. This project maintains its focus on the synergistic capabilities of breeding a plant and fish with the use of the same water.

Several government bodies are understanding that the significant roles for the state comprise of facilitating, coordinating and adopting reforms in order to enhance the business environment that is not directly associated to aquaculture but has spill-over influence on the sector. Some of the nations are divesting expensive infrastructure and undeliverable facilities. On the other hand, the others have adopted aquaculture-specific policies and developed framework strategies offering an idea and roadmap to direct the development. In addition to it, a few government bodies have offered soft credit lines in agricultural development and commercial banks, but access to reasonable credit, seed and feed of adequate quantity and quality, and land ownership or safe access to common property resources are main constraints to the advancement and/or intensification of production.

Figure 10.5: There are several species that are on the radar of the sustainable rural aquaculture, which will result in the rise in the numbers of these fishes, thus fulfilling the global demands.

Source: https://live.staticflickr.com/65535/48020421743_fd6639facd_b.jpg

The classification of species, selective breeding and the production of low-cost diets are the focal point of the research in a few centers. There are several nations that used the methods of on-farm participation in research. Also, the model farms and private enterprises are used which results in rapid diffusion of technologies through farmer-to-farmer pathways.

Usually, the extension services are weak and imperfectly resourced. There is an urgent requirement to enhance the individual services and also improve the links between research and development.

There is an increase in the involvement of the private sector in the production and delivery of inputs (seed and feeds) and the manufacture and supply of aquaculture equipment in some of the nations. At the same point of time, producer associations of one form or another are present in various nations at both the national as well as the local levels. These associations are playing a catalytic role in the sector, in aspects of flow of information, interchange of experience, and deciding the agenda and priority.

The establishment of operational clusters of farmers is contributing to effective delivery of support services, guaranteeing economies of scale, decreasing the transaction costs and enhancing the competitiveness. In other examples, grouping of farmers but particularly of farms has been a deliberate result of zoning regions for aquaculture that is based on the biophysical and socio-economic parameters of the specific site.

10.6. CASE STUDY: ECOSYSTEM AND PUBLIC HEALTH RISKS FROM NEARSHORE AND OFF-SHORE FINFISH AQUACULTURE

10.6.1. Introduction

Aquaculture is a diverse and growing food production sector that includes farming of fish, crustaceans, shellfish and aquatic plants in ponds, lakes, rivers, bays, estuaries and oceans. More than half of seafood consumed globally is now farmed, and aquaculture has surpassed global beef production in recent years. The United States (U.S.) is a net importer of seafood. In 2016, the U.S. imported edible seafood products valued at $19.5 billion, and exported $5 billion, leaving a $14.5 billion seafood trade deficit.

Domestic seafood production in the U.S. is skewed heavily toward wild-caught seafood. In 2013, the value of edible U.S. seafood production was $5.29 billion for wild-caught commercial landings and $1.15 billion for farmed products. Wild-caught seafood production is not expected to expand due to natural resource constraints, but there is growth potential for the aquaculture industry in the U.S. and abroad. There is interest among some aquaculture industry stakeholders and U.S. regulators in developing a near- and offshore finfish aquaculture (NOFA) industry in U.S. state and federal waters.

This form of aquaculture involves raising large numbers of finfish in net pens or cages near the water surface or in the water column. Farms could be located in the Atlantic or Pacific Oceans including Pacific Islands, the Gulf of Mexico, or the Great Lakes. These operations can be sited in open water similar to an oilrig or wind farm.

In this case study we summarize recent peer-reviewed scientific articles and reports that reflect the current state of commercial NOFA production in developed countries and impacts on aquatic ecosystems and humans. It is important for stakeholders to consider the environmental and public health implications when developing NOFA policies and regulations, and when investing in aquaculture operations.

10.6.2. Current Near- and Offshore Finfish Production in the U.S.

Commercial production of finfish in near-shore settings is limited in the U.S. to farmed Atlantic salmon in coastal Maine and Washington and farmed yellowtail in Hawaii. There are fewer than ten farm sites in Maine and Washington combined, all owned by Cooke Aquaculture, a vertically-integrated global aquaculture corporation.

Cooke Aquaculture is the 5th largest salmon producer in the world. Over the years, Cooke Aquaculture has consolidated the farmed salmon industry in the U.S. by purchasing their competitors, True North Salmon in Maine and Icicle Seafood in Washington. Maine and Washington are important for Cooke Aquaculture; these states represent roughly one fifth of their production (19,500 tons) valued at $77 million in 2015, but Maine and Washington are minor contributors (< 1%) to global farmed salmon production.

In Hawaii, a single company, Blue Ocean Mariculture, operates a net pen farm. In 2014, the company produced 450 tons of yellowtail in six net pens and planned to scale up to 1,100 tons in eight net pens by 2017. There has been some discussion of a net pen industry in the Great Lakes and state agencies have commissioned reports to explore this issue.

There is no commercial offshore finfish production in U.S. federal waters, however, the National Oceanic and Atmospheric Administration (NOAA) and the Gulf of Mexico Fisheries Management Council approved a permitting system for offshore finfish aquaculture in the region in 2016.

10.6.3. Fish Escapes

Farmed finfish are selectively bred over multiple generations to increase desirable traits like larger size, faster growth rates, or adaptation to captivity or breeding. Escapes of farmed fish remain a perennial issue for NOFA across multiple continents. Globally, several million fish escape net pen farms each year. In Europe, over a three-year period in the late 2000s, 242 incidents of escapes were recorded totaling over eight million escaped fish. In Canada, hundreds of thousands of fish escape net pen operations annually.

In August 2017, approximately 160,000 farmed Atlantic salmon escaped from net pens in Washington State. In this case, some research suggests there is low risk to wild salmon populations native to the Pacific coast, because recaptured fish were found to not be eating and were free of disease. However, fishers and environmentalists remain concerned about escapes.

Researchers have sampled water bodies looking for evidence of farmed fish. In one study, a team snorkeled in 41 wild salmon supporting rivers in Vancouver Island, Canada, and detected escaped farmed Atlantic salmon in over a third of those rivers. Other studies use long-lines to catch farmed fish in the wild. Once fish escape farms, the success rate of catching them and returning them to the farm are very low; averaging 8% across multiple studies.

Studies indicate there are both short-term and long-term ecological risks from escapes related to selective breeding and low genetic variability of farmed finfish. In the short-term, farmed fish can exert competitive pressure on native wild fish, as has been found with salmon. Scientists have studied the feeding habits of wild salmon and escaped farmed salmon in the North Atlantic by inspecting the stomach contents of caught fish. They found diets consumed by wild and escaped farmed fish were similar, suggesting that escaped farmed fish were well adapted to the wild.

Studies exploring long-term risks focus on genetic pollution and establishment of farmed fish populations in the wild. For example, farmed salmon populations exist in the wild in Chile and trace their ancestry to farmed broodstock from the U.S. and Canada.

Farmed Atlantic cod escapees were also found to reproduce in the wild. Established populations of farmed fish could provide an economic benefit to fishermen, but these benefits must be weighed against the ecological risks caused by escaped fish and the large economic loss caused by escapes. Researchers modeled the potential genetic impact of fish escapes from Gulf of Mexico net pen operations on wild cobia populations over 50 years.

They found that more escapes and use of genetically different source populations increased the genetic impacts on wild species. Proponents of NOFA argue that U.S. laws and regulations effectively address most of the potential environmental effects of NOFA, but accidental fish escapes similar to the recent escape in Washington are difficult to protect against and are likely to occur where NOFA sites operate.

10.6.4. Fish Waste

There is no mechanism to capture animal waste from NOFA, unlike terrestrial animal production where animal manure is ideally collected, composted, and used to build soil fertility. Fish waste instead deposits in sediment under cages and net pens or disperses into the water column and can travel outside the farm environment. Modelers have studied fish waste and estimated that in NOFA, every ton of fish produced results in an additional 69 kg of nitrogen and 10 kg of phosphorus released into the environment.

In 2010, NOFA production was estimated to be 5 million tons, resulting in an estimated 345 million kg of nitrogen and 50 million kg of phosphorus excreted in fish waste. Over the past four decades, feed conversion ratios in finfish aquaculture have been reduced and less nitrogen and phosphorus are released per unit production, but industry growth projections indicate significant increases in the amount of fish raised and therefore larger amounts of waste that can lead to nutrient pollution and eutrophication.

High nutrient levels in aquatic environments can cause algal blooms, resulting in low oxygen levels and mortality among aquatic animals. Several recent studies have been published about the impacts of uneaten fish feed and fish waste on the local environment and ecosystem, which adds to a growing body of literature on fish waste. Impacts can be separated into near-field impacts on sediments and the water column, and far-field impacts on the ecosystem.

Studies find a gradient of impacts with greater nutrients and waste proximate to the farm and diluted further from the farm. The distance waste travels varies based on animal stocking density, feeding rates, and siting issues such as water depth and water velocity or currents. Price et al. (2015) reviewed the impacts of marine cage culture on water quality and primary production of local biota (non-farm animals). The authors found nutrient enrichment in the water column within 100 meters of the farms (near-field effects) but not at greater distances, and these nutrients were consumed by local biota which agrees with other research.

The local nutrient impacts sometimes modifies the trophic structure of local biota, but the degree to which these impacts resulted in positive or negative outcomes was not reported. Far-field ecological effects can occur in intensively farmed regions over time. Impacts are variable and depend on farm siting, density of farms, and the strength of local regulations. Sarà et al. (2011) studied far-field effects of marine aquaculture by examining historic nitrogen and phosphorus loading.

They found nutrient and chlorophyll-a concentrations steadily increased in a Sicilian bay after marine finfish aquaculture operations were introduced and were at higher concentrations than water outside of the bay. In Norway, the risk of organic loading and eutrophication outside of the production area of the farm was considered low.

Ongoing monitoring of production areas is needed to ensure that near-field and far-field impacts are not damaging ecosystems and aquatic organisms. Integrated multi-trophic aquaculture (IMTA) is an emerging type of aquaculture that combines production of aquatic species that are fed (e.g., finfish, shrimp) and species that filter nutrients and particulates from the water column (e.g., bivalves, seaweed, sea cucumbers). The two primary benefits of the IMTA production model are reducing overall nutrient load in the area around a net pen or cage operation and diversifying income for aquaculture farmers.

10.6.5. Current State and Federal Policy

Nearshore operations in state waters are primarily regulated by the state agency in charge of implementing the National Pollutant Discharge Elimination System (NPDES) permits in the state (e.g., in Maine, it is the Department of Environmental Protection). NPDES permits are issued and managed to comply with the federal Clean Water Act, under authority granted by the Environmental Protection Agency. In general, state agencies issue location-based permits and collect information from producers on fish biomass, volumes of feed, use of therapeutants, and escapes.

Additional agencies are involved, and the level of transparency regarding producer data and resources devoted to regulating nearshore finfish aquaculture varies by state. NOAA is the lead federal agency in the U.S. developing and implementing regional permitting systems to allow NOFA in federal waters. The fishery management plan for offshore aquaculture in the Gulf of Mexico provides the first location where regulations and best management plans have been created and permits will be accepted for offshore aquaculture.

NOAA developed the regulatory structure for finfish production in the Gulf of Mexico based on the authority given to NOAA via the Magnuson-Stevens Fishery Conservation and Management Act (MSA). The MSA is the primary federal law regulating wild capture fisheries, and NOAA determined that finfish aquaculture in federal waters can be managed under the MSA. There is a pending legal action challenging inclusion of finfish aquaculture under the authority granted by the MSA.

NOAA has coordinated with the Environmental Protection Agency, U.S. Army Corp of Engineers, U.S. Coast Guard, Bureau of Ocean Energy

Management, and the Bureau of Safety and Environmental Enforcement to develop a permitting system in the Gulf of Mexico for offshore finfish production based on existing regulations.

Despite the involvement of these agencies, significant regulatory gaps remain at the state and federal levels, in part due to application of existing laws instead of development of a new regulatory system for NOFA. For example, in addition to the gaps regarding occupational health and safety described above, offshore operations may have an adverse impact on the aquatic ecosystem that is not subject to monitoring (i.e., does not result in higher mortality of a managed fish species), and therefore does not result in requirements to modify or stop production. In addition to leading policy formulation and regulation of aquaculture in federal waters, NOAA, as an agency within the Department of Commerce, has an explicit goal to promote and grow the U.S. aquaculture industry. This is similar to the U.S. Department of Agriculture, which regulates and promotes agriculture. To reduce the potential for a conflict of interest, NOAA should consider separating the roles of policy/regulatory development from industry promotion.

The current situation could lead to decisions regarding stringency of regulations and required levels of transparency that favor industry growth and profitability at the expense of protections for ecosystems and public health. A similar situation exists in Canada; an independent commission concluded that the Canadian Department of Fisheries and Oceans should not be responsible for regulating and promoting the farmed salmon industry.

10.6.6. Conclusion and Policy Recommendations

The case study focuses upon recent peer-reviewed literature relevant to potential impacts from NOFA in the U.S. Trade-offs between profitability and impacts to ecosystems and public health are a hallmark of many industries.

A robust, comprehensive regulatory system is necessary to ensure a potentially harmful industry consistently operates in a manner that minimizes risks, but the current regulatory structure for NOFA in the U.S. is inadequate. Due to the ongoing challenges associated with NOFA summarized above, current regulatory gaps, and potential impacts on public health and aquatic ecosystems, we recommend the following:

- Increase requirements for monitoring and reporting at offshore aquaculture sites to include monthly reports of disease outbreaks, therapeutant use, mortalities, escapes, and current feed use and fish biomass. All information should be posted by regulatory agencies on a website accessible to researchers and the public.

- Implement active environmental monitoring systems that test for therapeutants and breakdown products in fish tissue and sediment samples, fish pathogens, escaped farmed fish, nutrient loading, and antibiotic resistant bacteria in sediments. The monitoring system should be fulfilled by trained agency staff, not industry staff.

- Develop a robust set of requirements to protect and monitor the health and safety of NOFA workers. Data on injuries, illnesses, and deaths should be reported to the Occupational Safety and Health Administration. Information specific to each NOFA operation should be posted on a website accessible to researchers and the public.

- Separate federal and state policy/ regulatory efforts from NOFA industry promotion to reduce potential conflicts of interest.

- Until the above recommendations are fully implemented, do not approve operations in the Gulf of Mexico and do not implement new permitting systems in other regions of the U.S.

REFERENCES

1. An evaluation of small scale Freshwater Rural Aquaculture Development of Poverty Reduction. (2019). [ebook] Asian Development Bank. Available at: https://www.adb.org/sites/default/files/publication/27961/fresh-water.pdf [Accessed 29 August 2019].

2. Anderson, J., Asche, F., Garlock, T., & Chu, J. (2017). Aquaculture: Its Role in the Future of Food. *Frontiers of Economics and Globalization*, [online] pp.159–173. Available at: https://www.emerald.com/insight/content/doi/10.1108/S1574–87152017000017011/full/html [Accessed 29 August 2019].

3. Asche, F., Roll, K., & Tveterås, S. (2008). Future Trends in Aquaculture: Productivity Growth and Increased Production. *Aquaculture in the Ecosystem*, [online] pp.271–292. Available at: https://link.springer.com/chapter/10.1007/978–1–4020–6810–2_9 [Accessed 29 August 2019].

4. Belton, B. (2017). *Myanmar Aquaculture Drives Development, Shows Growth Potential – WorldFish Blog*. [online] WorldFish Blog. Available at: http://blog.worldfishcenter.org/2017/04/myanmar-aquaculture-drives-development-shows-growth-potential/ [Accessed 29 August 2019].

5. Changing the Face of the Waters. The Promise and Challenge of Sustainable Aquaculture. (2007). [ebook] International Bank for Reconstruction and Development/The World Bank. Available at: http://documents.worldbank.org/curated/en/322181468157766884/pdf/416940PAPER0Fa18082137015501PUBLIC1.pdf [Accessed 29 August 2019].

6. Diana, J. (2009). Aquaculture Production and Biodiversity Conservation. *BioScience*, [online] *59*(1), 27–38. Available at: https://academic.oup.com/bioscience/article/59/1/27/306930 [Accessed 29 August 2019].

7. Fry, J., Love, D., & Innes, G. (2018). Ecosystem And Public Health Risks From Nearshore And Offshore Finfish Aquaculture. [PDF] Available at: https://www.jhsph.edu/research/centers-and-institutes/johns-hopkins-center-for-a-livable-future/_pdf/research/clf_reports/offshor-finfish-final.pdf [Accessed 30 Jul. 2019].

8. Funge-Smith, S., Moehl, J., & Halwart, M. (2019). *The Role of Aquaculture in Rural Development*. [ebook] FAO Fisheries Department.

Available at: http://www.fao.org/tempref/docrep/fao/005/Y4490e/y4490e.pdf#page=50 [Accessed 29 August 2019].

9.   Giriappa, S. (1994). *Role of Fisheries in Rural Development*. Delhi: Daya Pub. House.

10.   Haylor, G., & Bland, S. (2019). *Integrating Aquaculture into Rural Development in Coastal and Inland Areas*. [online] Fao.org. Available at: http://www.fao.org/3/ab412e/ab412e31.htm [Accessed 29 August 2019].

11.   Jaiswal, K. (2017). *Role of Aquaculture in Rural Development*. [online] Slideshare.net. Available at: https://www.slideshare.net/KRISHNACOF/role-of-aquaculture-in-rural-development [Accessed 29 August 2019].

12.   Jana, B., & Jana, S. (2003). The Potential and Sustainability of Aquaculture in India. *Journal of Applied Aquaculture*, [online] 13(3–4), 283–316. Available at: https://www.tandfonline.com/doi/abs/10.1300/J028v13n03_05 [Accessed 29 August 2019].

13.   Lehane, S. (2013). *Fish for the Future: Aquaculture and Food Security*. Nedlands: Future Directions International Pty Ltd.

14.   Muir, J. (2005). Managing to harvest? Perspectives on the potential of aquaculture. *Philosophical Transactions of the Royal Society B: Biological Sciences*, [online] 360(1453), 191–218. Available at: https://www.ncbi.nlm.nih.gov/pmc/articles/PMC1636102/ [Accessed 29 August 2019].

15.   Philipps, R. (2012). *International Aquaculture: Past, Present, and Future Case Studies: Cambodian Sustainable Rice Fish Integration & Namibian Aquaculture*. [ebook] UNIVERSITY OF WISCONSIN-STEVENS POINT. Available at: https://www.uwsp.edu/forestry/StuJournals/Documents/IRM/International%20Aquaculture%20Past%20Present%20and%20Future%20-%20Rebecca%20Phillips.pdf [Accessed 29 August 2019].

16.   Poppick, L. (2018). *The Future of Fish Farming May Be Indoors*. [online] Scientific American. Available at: https://www.scientificamerican.com/article/the-future-of-fish-farming-may-be-indoors/ [Accessed 29 August 2019].

17.   Rajee, O., & Kar Mun, A. (2017). *Impact of Aquaculture on the Livelihoods and Food Security of Rural Communities Olaganathan Rajee and Alicia Tang Kar Mun*. [ebook] International Journal of

Fisheries and Aquatic Studies. Available at: http://www.fisheriesjournal. com/archives/2017/vol5issue2/PartD/5–1-81–919.pdf [Accessed 29 August 2019].

18. Satia, B. (2010). *Aquaculture Development in Africa: Current Status and Future Prospects*. [online] Network of Aquaculture Centers in Asia-Pacific. Available at: https://enaca.org/?id=170 [Accessed 29 August 2019].

19. Sharma, A., & Saxena, D. (2019). *Aquaculture's Hope for Bright Future for Rural People*. [online] Aquafind.com. Available at: http://aquafind. com/articles/Aquaculture-Hope.php [Accessed 29 August 2019].

20. Tsani, S., & Koundouri, P. (2018). A Methodological Note for the Development of Integrated Aquaculture Production Models. *Frontiers in Marine Science*, [online] 4. Available at: https://www.frontiersin.org/ articles/10.3389/fmars.2017.00406/full [Accessed 29 August 2019].

21. Turchini, G., Trushenski, J., & Glencross, B. (2018). Thoughts for the Future of Aquaculture Nutrition: Realigning Perspectives to Reflect Contemporary Issues Related to Judicious Use of Marine Resources in Aquafeeds. *North American Journal of Aquaculture*, [online] *81*(1), 13–39. Available at: https://afspubs.onlinelibrary.wiley.com/ doi/10.1002/naaq.10067 [Accessed 29 August 2019].

22. Waite, R. (2014). *Sustainable Fish Farming: 5 Strategies to Get Aquaculture Growth Right*. [online] World Resources Institute. Available at: https://www.wri.org/blog/2014/06/sustainable-fish-farming-5-strategies-get-aquaculture-growth-right [Accessed 29 August 2019].

23. Worldfishcenter.org. (2019). *The Importance of Fish | WorldFish Organization*. [online] Available at: https://www.worldfishcenter.org/ why-fish [Accessed 29 August 2019].

24. Worldfishing.net. (2019). *World Fishing & Aquaculture | Canada Focuses on the Future*. [online] Available at: https://www.worldfishing. net/news101/regional-focus/canada-focuses-on-the-future [Accessed 29 August 2019].

INDEX

Cultivation 20

D

Denitrification system 52
Dietary animal protein 91
Direct current (DC) 61
Disease resistance   102, 109, 110, 111
Disease susceptibility 109
Dissolved inorganic nutrients 126
Domestic systems 27

E

Early Mortality Syndrome (EMS) 142
Economic development 8
Egg production 59
Employee-oriented corporate approach 74
Environmental contamination 55
Environmental degradation 89
Environmental pollution 19, 25
Extension production shelf-life 56

F

Fair management-labor relationship 74
Farming community 11, 13
Farming oysters 7
Farm sustainability 7, 92, 97
Fertilization 69
Fish bio diversity 34
Fish farming 8
Fish production 34, 35, 36
Fish production program 127
Floating farm 58
Follicle stimulating hormone (FSH) 107
Food and Agriculture Organization of the United Nations (FAO) 148
Food security 88, 91, 93, 94, 95, 100
Freshwater production system 37
Full- cycle technology 73

G

Galvanotaxis 61
Gene banking 104, 105, 106
Genetic engineering 102, 103, 105, 106, 113, 115
Global-level monitoring 77
Gonadotropin Releasing Hormone (GnRH) 107, 108
Gross Domestic Product (GDP) 128

H

Hatchery operator 71, 73
Health management 105, 106
Herbivorous 38
Herbivorous milkfish 4
High density intensive feeding system 41
Horticulture 39
Human intervention 121
Humid climate 4

I

Immune system 111
Induced breeding 104, 105, 106, 108
Industrial farming 19
Industrial production systems 21
Infectious Salmon Anemia (ISA) 60
Inland aquaculture 127, 130
Inorganic fertilizers 7, 24
Integrated Multi-Trophic Aquaculture (IMTA) 126
Integrated pest management 29